Abdelali FATEH

Challenges and issues of urban development in Moroccan cities

Abdelali FATEH

Challenges and issues of urban development in Moroccan cities

Towards socio-economic equity and spatial integration

ScienciaScripts

Contents

Preface 3

General introduction 4

Part I 7

Part II 46

Part III 83

General conclusion 106

Coordinated by :
Abdelali FATEH

Contributors :

Abdelali FATEH, *Mohammed V University, Rabat - Morocco*

Mohamed KAFSSI*, Faculty of Education, Mohammed V University, Rabat - Morocco*

Anas EL AZIZI EL ALAOUI, *Faculty of Arts and Sciences Humaines, Mohammed V University, Rabat - Morocco*

Soukaina BOUJEMI*, Faculty of Arts and Sciences Humaines, Mohammed V University, Rabat - Morocco*

Malak JABLI*, Faculty of Humanities, Mohammed V University, Rabat - Morocco*

Rachida BAILI*, Doctor in Geo-environment Member of the Geo-engineering and Environment Laboratory (LGIE) Faculty of Science, Meknes.*

Jamila OUARDI*, Centre regional des metiers de l'education et de formation (CRMEF, Casablanca -Settat).*

Moroccan cities have been experiencing rapid urban growth for several decades, leading to numerous challenges in terms of land use planning, resource management and socio-economic integration. Faced with these challenges, the need for sustainable urban development has become an inescapable imperative to ensure the sustainability and well-being of city dwellers.

This collective work brings together the reflections and analyses of researchers and doctoral student-researchers, highlighting the main issues relating to sustainable urban development in Morocco. It addresses crucial issues such as the functional integration of urban spaces, strategic urban planning, education in sustainable urban planning, the preservation of architectural heritage and urban mixing. Particular attention is also paid to the issues of equity, socio-spatial justice and the role of urban intelligence in reducing territorial disparities.

By offering a multidisciplinary perspective and a holistic approach, this book is intended to be a valuable tool for understanding the complex challenges facing Moroccan cities and identifying the levers of action for more inclusive, resilient and environmentally-friendly urban development.

The contributions gathered here bear witness to the commitment and expertise of researchers, urban planners and architects working for a sustainable urban future in Morocco.

They invite us to rethink our models of urban development and adopt innovative practices, while putting citizens at the heart of this reflection.

We hope that this book will stimulate a fruitful dialogue between the various players involved and contribute to the emergence of ambitious visions and policies to meet the challenges of sustainable urban development in Moroccan cities.

Abdelali FATEH

The phenomenon of urbanisation is accelerating at a staggering rate worldwide, leading to profound changes in urban landscapes. According to United Nations projections, almost 70% of the world's population will live in cities by 2050.

This exponential urban growth is putting considerable pressure on natural resources and the environment. Cities face major challenges such as air pollution, waste management, road congestion and biodiversity loss. These complex environmental issues threaten the quality of life of city dwellers and the sustainability of urban ecosystems.

Faced with these major challenges, it is imperative that we rethink our traditional methods of urban development. Current models, often based on rapid economic growth at the expense of environmental protection, are proving to be unsuitable and unsustainable in the long term. It is now crucial to adopt a holistic, integrated approach that takes account of the environmental, social and economic dimensions of urban development. This means challenging our current urban paradigms and exploring new, more sustainable and resilient avenues.

The concept of sustainable urban development is emerging as a crucial response to these complex challenges. It aims to create cities that are more environmentally friendly, socially inclusive and economically viable. Harmonious development means using

In addition, it requires the active participation of citizens and the consideration of social aspects such as equitable access to essential services and quality of life. It also requires the active participation of citizens and the consideration of social aspects such as equitable access to essential services and quality of life. By adopting a sustainable and holistic approach, cities can become healthier, more resilient and attractive places for present and future generations.

The aim of this multidisciplinary collective publication is to explore the many dimensions of sustainable urbanism, offering both theoretical and practical insights. Using a cross-disciplinary approach, this publication brings together contributions from researchers from a range of disciplines including urban planning, architecture, geography, sociology and economics. This range of perspectives makes it possible to tackle the complex issues of sustainable urban development from a number of complementary angles.

The authors analyse a range of issues, from the functional integration of urban spaces to sustainable strategic planning, including education, heritage, architectural mix, socio-spatial equity and urban intelligence. This multidisciplinary approach aims to provide an in-depth understanding of the challenges and opportunities associated with implementing sustainable urbanism in different contexts.

Although the principles of sustainable urban planning are universal, their application must take account of local and regional specificities. For this reason, this publication pays particular attention to the Moroccan context, which is rich in its own specificities.

urban history and its current challenges in terms of sustainable urban development.

The case studies presented, such as the conurbations of Rabat-Sale-Temara and Casablanca, as well as the analysis of Moroccan coastal cities, enable the theoretical reflections to be anchored in concrete realities. This contextualisation provides an opportunity to explore the local specificities, constraints and opportunities specific to Morocco, while drawing out lessons relevant to other regions facing similar challenges.

By exploring these various facets of sustainable urban planning in the Moroccan context, this publication aims to stimulate dialogue and encourage the exchange of experience and good practice between the various players involved in promoting sustainable urban development in Morocco and beyond.

Part I

Functional integration and strategic planning of urban areas

The functional integration of different urban areas within a single economic metropolis represents a major challenge for urban planners and political decision-makers. The rapid growth of cities and the sprawl of peri-urban areas have led to a spatial fragmentation that is detrimental to the coherence and efficiency of urban systems. Faced with this problem, a strategic approach and an overall vision are essential to ensure genuine synergy between the various components of the metropolis and thus optimise the flow of people, goods and information.

From this point of view, the Rabat-Sale-Temara urban conurbation is a particularly interesting case study, as it illustrates the issues involved in the functional integration of urban areas that were once distinct but are now intertwined in a single metropolitan dynamic. An in-depth analysis of this conurbation will shed light on the specific challenges facing Moroccan cities in their quest for greater territorial cohesion and better articulation of urban functions.

At the same time, the close relationship between sustainable strategic planning and the urban competitiveness of Moroccan coastal cities merits particular attention. In a context of increasing globalisation and fierce competition between metropolises, a city's ability to stand out and attract investment and talent depends largely on its capacity to project itself into the future in a sustainable manner and to make the most of its specific assets.

Functional integration between urban areas in the context of the political and economic metropolis: the case of the Rabat-Sale-Temara urban agglomeration

Abdelali FATEH [1]

Resume :

In a context of increasing monopolisation and integration of urban spaces, this research looks at the dynamics of functional interdependence at work within metropolitan agglomerations.

Drawing on different theoretical approaches to functional urban regions and complex urban systems, the article analyses the processes and forms of integration at play in the specific case of the Rabat-Sale-Temara conurbation in Morocco. It examines the historical dynamics of spatial integration, mobility infrastructures, intersectoral economic complementarities, the distribution of metropolitan facilities and services, and governance issues at the supra-municipal level.

The aim of this article is to shed some theoretical light on the functional interdependencies that structure contemporary metropolises. [1]

Key words: Functional integration, Urban areas, Metropolis, Functional urban regions, Urban systems, Rabat-Sale-Temara, Interdependencies, Mobility, Economic complementarities, Metropolitan governance.

Introduction

In a context of increasing metropolisation and the reshaping of urban dynamics on a global scale, the functional integration of urban areas within metropolitan regions is a major challenge for both research and public action.

This issue is at the heart of contemporary debates on changes in urban systems and the emergence of new forms of polycentric centrality. The aim of this research is to analyse the processes and forms of functional integration at work in the Rabat-Sale-Temara conurbation, using the theoretical frameworks of functional urban regions and the systemic approach to urban spaces. The aim is to examine how economic interdependencies, mobility flows, the distribution of metropolitan facilities and governance mechanisms contribute to structuring a functionally integrated urban space that goes beyond administrative boundaries.

The article thus aims to shed empirical and theoretical light on the dynamics of functional integration that characterise contemporary metropolises, particularly in the context of emerging countries. The relevance of this subject lies in its ability to contribute to reflections on the issues of planning and governance of metropolitan regions, with a view to sustainable urban development and strengthening territorial

---

[1] *Mohammed V University, Rabat*

competitiveness.

I. Theoretical approaches to the functional integration of urban spaces

I.1 The theory of functional urban regions

1.1.1 Definitions and key concepts

The theory of functional urban regions (FURs) has established itself as a major conceptual framework for understanding the dynamics of spatial and economic integration in contemporary metropolitan areas. Initially developed in the 1970s by Brian Berry and his colleagues, this approach aims to go beyond the limits of traditional administrative divisions in order to grasp the real extent of the functional interactions structuring urban agglomerations.

RUFs are defined as coherent territorial units characterised by a high intensity of flows and exchanges between a central urban centre and its peripheral zone of influence. This definition is based on the concept of the employment area, focusing on commuting as a key indicator of functional integration. Beyond this initial dimension, the notion of RUF has gradually been broadened to encompass other types of interaction, such as consumer flows, business networks and leisure and education-related mobility.

1.1.2 Integration analysis criteria

The analysis of RUFs is based on a number of criteria for assessing the degree of functional integration of urban areas. These criteria include, first and foremost, the intensity and direction of daily travel flows, which reveal the polycentric or monocentric structure of the conurbation (Ludovic Halbert, 2021, p. 78).

Economic complementarity between the various urban centres is a second essential criterion, highlighting functional specialisations and productive synergies on a metropolitan scale.

The spatial distribution of facilities and services with a metropolitan impact represents a third area of analysis, making it possible to identify the structuring centralities and hierarchical patterns within the RUF.

Finally, the existence of governance and planning mechanisms at supra-municipal level is a fourth criterion, revealing the institutional awareness of functional interdependencies (Gilles Pinson and Christophe Demaziere, 2017, p. 89).

By combining these different criteria, we can understand the complexity of the processes of integration of the Louvre into functional urban regions, going beyond a purely morphological approach to grasp the socio-economic and political dynamics that shape these metropolitan areas.

I.2 The concept of the metropolis

1.2.1 Emergence and characteristics

The concept of the metropolis has gradually emerged in the field of urban studies to designate the major conurbations that play a driving role in the contemporary global economy. The emergence of this concept is closely linked to the processes of globalisation and metropolisation that have profoundly reconfigured the urban hierarchy since the 1980s (Pierre Veltz, 2019, p. 23).

Economic metropolises are characterised above all by their ability to concentrate and articulate strategic economic functions on a global scale, such as advanced business services, research and development activities and creative industries (Ludovic Halbert, 2018, p. 56).

This concentration is generally accompanied by strong demographic appeal, pronounced social and cultural diversity, and integration into dense international exchange networks (Saskia Sassen, 2021, p. 112). The economic metropolis is thus distinguished from mere large cities by its extraterritorial influence and its ability to polarise the flow of capital, information and talent on a supra-regional or even global scale.

1.2.2 Role in structuring economic activities

The structuring role of economic activities in the formation and development of metropolises is a central element of this concept. The concentration of higher-level economic functions generates agglomeration effects and positive externalities that enhance the attractiveness and competitiveness of the metropolitan area.

These dynamics are reflected in a spatial reconfiguration of activities within the metropolitan area, with the emergence of specialised poles and innovative clusters that structure the local economic fabric.

Furthermore, economic metropolises are characterised by a high degree of inter-sectoral integration, favouring synergies between traditional industries, advanced services and the knowledge economy. This integration is particularly evident in the intensification of subcontracting and partnership relations between companies, as well as in the multiplication of interactions between the productive sphere and the world of research.

Finally, the economic dynamism of metropolises has a knock-on effect on the whole of their region of influence, through mechanisms for redistribution and the spread of innovation that help to structure coherent regional production systems.

I.3 The systemic approach to urban spaces

1.3.1 The city as a complex system

The systemic approach to urban spaces has become an essential paradigm for understanding the complexity of contemporary metropolitan dynamics. This theoretical perspective, initially developed in the 1960s by researchers such as Brian Berry and Jay Forrester, sees the city as a complex system, characterised by a multitude of constantly interacting elements.

From this perspective, the urban space is analysed as an organic whole, whose components are interdependent and influence each other through positive and negative feedback loops (Denise Pumain and Lena Sanders, 2020, p. 67). This approach enables us to move beyond a static and compartmentalised view of the city and grasp the dynamic processes that underpin its functioning and evolution. It highlights the emerging properties of the urban system, which cannot be reduced to the simple sum of its parts, but result from the complex interactions between its different components.

1.3.2 Functional interactions and interdependencies

Functional interactions and interdependencies are at the heart of the systemic analysis of urban spaces. These multifaceted relationships are expressed through the various flows and exchanges that structure the day-to-day functioning of the metropolis: flows of people, goods, information, capital, and so on.

The study of these interactions highlights the functional complementarities between the various sub-assemblies of the urban system, whether neighbourhoods, municipalities or specialised activity centres. It also reveals the hierarchies and asymmetries that characterise these relationships, highlighting the phenomena of polarisation and dependence within the metropolitan area.

Furthermore, the systemic approach seeks to identify the feedback loops that amplify or regulate these interdependencies, thereby contributing to the stability or transformation of the urban system as a whole (Lena Sanders and Denise Pumain, 2023, p. 201). In particular, this perspective makes it possible to analyse the cascading effects that local interventions can have on the metropolitan system as a whole, thus providing a relevant framework for understanding urban planning and governance issues on the scale of vast, functionally integrated territories.

II. Integration dynamics at work in the Rabat-Sale-T emara conurbation

II.1 Elements of context

2.1.1 Geographical location and demographic weight

The Rabat-Sale-Temara conurbation is a major urban area in Morocco, whose geographical location and demographic weight make it a particularly relevant case study for analysing the dynamics of the functional integration of metropolitan areas. Located on the country's Atlantic coast, around 90 km south of Casablanca, this conurbation covers an area of around 1,800 km^2 , including the cities of Rabat (the administrative capital), Sale and Temara.

Map (01) : The Rabat - Sale - Temara conurbation

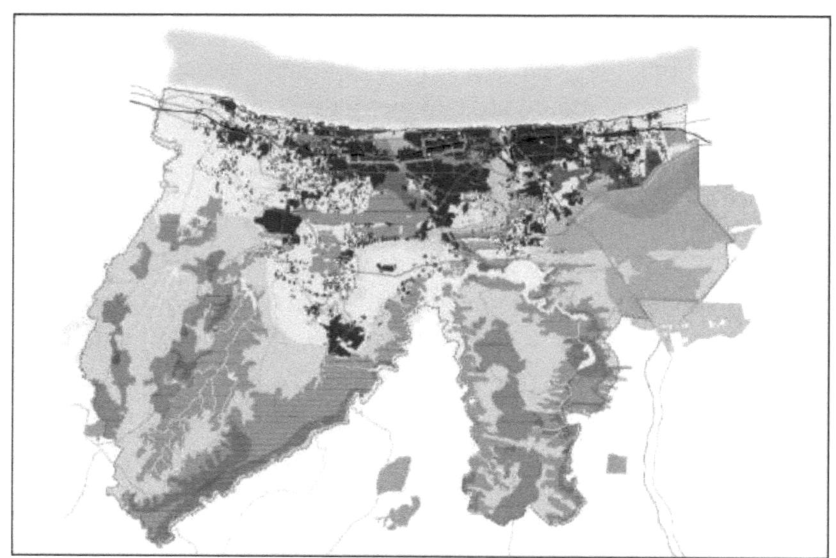

According to the latest demographic estimates, the conurbation has a population of almost 2.5 million, making it the country's second largest urban concentration after Greater Casablanca (Haut Commissariat au Plan du Maroc, 2022, p. 23). This sustained demographic growth, with an average annual rate of 2.1% over the period 2014-2022, bears witness to the continuing attractiveness of the capital region, particularly for rural populations seeking economic opportunities.

Figure (01) : Demographic trends in the Rabat-Sale-Temara conurbation between 1994 and 2022

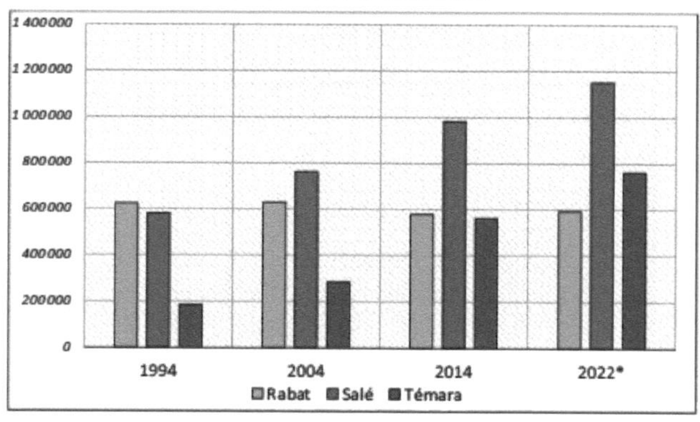

14

Estimated for 2022

Sources :
- General population and housing censuses (RGPH) of 1994, 2004 and 2014 (Haut-Commissariat au Plan du Maroc)
- Estimates for 2022 based on demographic projections by the Moroccan High Commission for Planning

However, the spatial distribution of this population reveals major disparities between the different parts of the conurbation, with Sale and Temara becoming denser, while Rabat is experiencing relative demographic stabilisation due to the saturation of its urban fabric.

2.1.2 National and regional economic importance

Economically, the Rabat-Sale-Temara conurbation plays a key role in Morocco's national and regional economy. As the kingdom's administrative capital, Rabat concentrates the bulk of government and diplomatic functions, generating a large tertiary sector linked to public service activities.

In addition to this administrative specialisation, the conurbation has undergone diversified economic development over the last few decades, with the emergence of business clusters in high-value-added sectors such as information and communication technologies, offshoring and creative industries. According to data from the Moroccan Statistics Directorate, the conurbation contributes 16.5% to the national GDP, making it the country's second largest economic centre (Moroccan Statistics Directorate, 2023, p. 34).

On a regional scale, Rabat-Sale-Temara has a significant knock-on effect on the surrounding areas, particularly through the development of peripheral industrial and logistics zones such as Ain Aouda. However, this economic dynamism is accompanied by major challenges in terms of territorial balance and social cohesion, with persistent disparities between the different parts of the conurbation in terms of access to employment and urban services.

II.2 Historical processes of urbanisation and spatial convergence

2.2.1 Urban development and centrifugal dynamics

The urban development of this metropolitan area has been characterised by particularly pronounced centrifugal dynamics since the second half of the 20th century. The gradual saturation of Rabat's urban fabric, combined with growing land pressure, led to urban sprawl towards the outskirts (FATEH A, 2023), particularly towards Sale and Temara. This phenomenon has resulted in the emergence of new residential areas, often in the form of large housing estates or periurban housing estates (FATEH A 2023), which have gradually filled in the interstitial spaces between the historic urban cores. This centrifugal dynamic has been strongly influenced by public housing and development policies, which have favoured the horizontal extension of the city rather than the densification of the centre, as shown by the development of vast property developments such as Hay Riad in Rabat, Al Boustane in Sale and the Al Faourate zone in Temara.

Alongside these centrifugal dynamics, there has been a gradual process of conurbation between Rabat, Sale and Temara, which has led to the formation of an increasingly integrated urban continuum. This process of spatial convergence accelerated from the 1980s onwards, under the combined effect of demographic growth, the rural exodus and economic change (Fateh A. 2023). The conurbation materialised through the densification of the communication routes linking the three towns, particularly along the Bouregreg valley and the Atlantic coast. The construction of the Hassan II bridge in 2011, directly linking Rabat and Sale, marked an important symbolic and functional step in this process of spatial integration.

2.2.2 The phenomena of conurbation and metropolisation

The phenomenon of metropolisation, for its part, has become more recent, reflecting the transition from a simple conurbation to a functionally integrated metropolitan area. This dynamic is characterised by a redistribution of urban functions across the conurbation, with the emergence of new peripheral centres and increased specialisation of the various urban centres. While Rabat retains its primacy in terms of administrative and cultural functions, Sale and Temara have developed significant centres of economic activity, particularly in the industrial and logistics sectors (Ain Atig industrial zone (Skhirat-Temara Province). The Bouregreg valley development project, launched in 2006, illustrates this desire to create a coherent metropolitan area, by creating new links between the riverbanks and developing regional facilities.

However, this metropolisation is accompanied by major challenges in terms of governance and territorial cohesion, with the administrative fragmentation inherited from history making it difficult to implement coordinated planning policies on a conurbation-wide scale. Socio-spatial disparities persist between the different parts of the metropolis, reflecting unequal access to urban resources and residential segregation.

II.3 Communication infrastructure and commuting 2.3.1 Core transport networks

The conurbation's main transport networks have undergone significant development over the last few decades, with the aim of improving connectivity both inside and outside the city. The conurbation's transport system is built around a number of major infrastructures:

a) The road network, dominated by the A1 motorway linking Rabat to Casablanca, and the A2 motorway to Fes, acts as a backbone for metropolitan and inter-urban travel.

b) The rail network, with Rabat-Ville station as the central hub, connecting the conurbation to the rest of the country via high-speed (LGV) and conventional lines.

c) The Rabat-Sale tramway, which came into service in 2011, is the backbone of metropolitan public transport, with two lines totalling 27 km and carrying more than 110,000 passengers every day (Societe du Tramway de Rabat-Sale, 2022, p. 12).

d) The urban and interurban bus network, which completes the coverage of the region, although its effectiveness is often questioned.

e) Infrastructure to cross the Bouregreg, notably the Hassan II bridge (2011) and the Moulay Hassan bridge (2016), which have considerably improved links between Rabat and Sale.

Despite these advances, there are still disparities between the different parts of the conurbation, with a concentration of heavy infrastructure in the central areas to the detriment of the expanding suburbs.

2.3.2 Daily travel flows

Daily travel flows within the Rabat-Sale-Temara conurbation bear witness to the intensity of functional interactions between the various urban centres. According to a household-travel survey carried out in 2020 (Agence Urbaine de Rabat-Sale, 2021, p. 78), the following trends can be observed:

a)	A total volume of around 4.5 million daily journeys within the conurbation, of which almost 60% are commuter journeys between home and work or home and study.

b)	A strong polarisation of flows towards Rabat, which concentrates most of the administrative and higher tertiary jobs. More than 200,000 people living in Sale and 100,000 in Temara commute to Rabat every day to work.

c)	An increase in transverse flows between Sale and Temara, reflecting the emergence of new economic centres in these municipalities.

d)	The car predominates in the modal split (45% of journeys), followed by walking (30%), public transport (20%, including 7% for the tramway) and two-wheelers (5%).

These mobility flows highlight the growing interdependence between the different parts of the conurbation, while also highlighting the challenges in terms of road congestion and saturation of public transport at peak times.

Managing these commuter flows is a major challenge for metropolitan planning. Local authorities are faced with the need to develop an integrated approach to mobility at the scale of the conurbation, linking transport policies and land-use strategies. Projects such as the extension of the tramway network, the creation of Bus Rapid Transit (BRT) lines and the development of multimodal interchanges are currently being studied to improve the fluidity of travel and strengthen the territorial cohesion of the metropolis.

## 111.	Forms of integration and functional interdependencies

3.1 Cross-sectoral economic complementarities 3.1.1 Employment areas and labour markets

In terms of employment areas and labour markets, the conurbation has a polycentric structure marked by distinct territorial specialisations. There are three main employment areas:

a)	Rabat, which concentrates the bulk of administrative jobs and higher metropolitan functions, particularly in the sectors of public governance, research and higher education.

b)	Sale, characterised by a more diversified economic fabric, combining traditional industrial activities (textiles, agri-food) with the emergence of new sectors such as information and communication technologies (ICT).

c)	Temara is a fast-growing industrial and logistics hub, home to a number of business parks.

This spatial distribution of jobs generates significant flows of labour between the different parts of the conurbation. According to a recent study by the Agence Urbaine de Rabat-Sale (2022, p. 56), almost 40% of working people living in Sale and 35% of those living in Temara work in Rabat on a daily basis, demonstrating the growing interdependence of the local labour markets.

At the same time, there is a trend towards the functional specialisation of the various urban centres, which is helping to strengthen economic complementarities on a metropolitan scale. This is illustrated by the emergence of integrated economic clusters, such as the automotive complex around the Renault-Nissan plant in Tangier, which is having a knock-on effect on the whole of the Rabat-Sale-Temara conurbation through the development of subcontractors and related services.

3.1.2 Productive activities and tertiary services

In terms of productive activities and tertiary services, the Rabat-Sale-Temara conurbation presents a diversified economic panorama, characterised by strong tertiarisation but also by the maintenance of a significant industrial fabric. According to data from the Direction de la Statistique du Maroc (2023, p. 34), the sectoral breakdown of employment in the conurbation is as follows:

- Services: 65% (including 25% for public administration)
- Industry: 20%.
- Construction: 10% of total
- Agriculture: 5% of sales

This economic structure reflects the coexistence of several dynamics:

a) A concentration of high value-added services in Rabat, particularly in the fields of finance, business consultancy and research and development. The Technopolis project, launched in 2008, illustrates this desire to position the city as a hub for innovation and advanced services. The maintenance of a diversified industrial fabric, particularly in Sale and Temara, with specialisations in food processing, textiles, mechanical engineering and, more recently, electronics and automobiles. The Ain El Aouda industrial estate has

Temara is an emblematic example of this productive dynamic.

b) The growth of the digital and creative economy, reflected in the development of specialist clusters such as the Rabat Technopark, fostering synergies between start-ups, major companies and research institutions.

These intersectoral complementarities are reflected in an intensification of economic exchanges within the conurbation. Subcontracting and partnerships between companies located in different parts of the metropolitan area are on the increase, helping to strengthen the functional integration of the economic fabric.

However, these dynamics are also accompanied by major challenges in terms of territorial cohesion and the balanced distribution of economic activities. The persistent concentration of skilled jobs in Rabat raises questions about the ability of other urban centres to retain and attract talent, while the redevelopment of brownfield sites in central areas raises issues of urban renewal and functional mix.

3.2 Metropolitan facilities and services

3.2.1 Educational, health and recreational facilities...

The Rabat-Sale-Temara conurbation is characterised by a significant concentration of metropolitan infrastructure, albeit with an uneven spatial distribution that reflects historical legacies and recent development dynamics.

In the field of education, Rabat is asserting its pre-eminence as a major national university centre. The Mohammed V University, spread over several campuses, is home to more than 80,000 students and has an influence well beyond the city limits (Moroccan Ministry of Higher Education, 2022, p. 45). This concentration of higher education establishments is reflected in major commuter flows of students from Sale and Temara, reinforcing the functional integration of the metropolitan area.

As far as health infrastructure is concerned, the conurbation benefits from a relatively dense network, although state-of-the-art facilities are predominantly located in Rabat. The Centre Hospitalier Universitaire Ibn Sina and the Hopital Cheikh Zayd are national references, attracting patients from all over the region (Moroccan Ministry of Health, 2023, p. 78).

However, efforts are underway to redress the balance, with new hospitals being built in Sale and Temara to improve access to healthcare across the metropolitan area.

Cultural and leisure facilities of metropolitan importance are also mainly concentrated in Rabat, with emblematic infrastructures such as the Mohammed V National Theatre, the Mohammed VI Museum of Modern and Contemporary Art, and the Prince Moulay Abdellah sports complex. However, recent projects show a desire to diversify the cultural and leisure offer throughout the city.

3.2.2 Commercial centres and centres of attraction

The commercial structure of the Rabat-Sale-Temara conurbation has undergone profound changes over the last two decades, with the emergence of new centres that are redrawing the geography of consumer flows on a metropolitan scale.

Rabat city centre is still very attractive, particularly around traditional shopping streets such as Avenue Mohammed V. However, there is a trend towards a proliferation of peripheral shopping centres, which are emerging as genuine metropolitan centres. For example, the Morocco Mall Rabat mega-shopping centre, soon to open on the boundary between Rabat and Temara, is generating consumer flows from the whole of the conurbation and beyond.

In Sale, the development of the marina and the Bab Al Bahr project has contributed to the emergence of a new commercial and leisure centre, reinforcing the town's attractiveness within the conurbation.

In Temara, the commercial area around Avenue Hassan II is gradually establishing itself as a major magnet for people living in the fast-growing outlying districts.

This reconfiguration of commercial centres is accompanied by changes in consumption and leisure practices on a metropolitan scale. As shown by a survey carried out by the Agence Urbaine de Rabat-Sale (2022, p. 112), almost 40% of Sale and Temara residents regularly visit Rabat's major shopping centres, while 25% of Rbatis visit the new shopping areas in Sale or Temara at least once a month.

However, this proliferation of centres of attraction raises issues in terms of mobility and accessibility. Road congestion on the outskirts of major shopping centres and the saturation of public transport at peak times demonstrate the need for better coordination between commercial planning and transport policy on a metropolitan scale.

3.3 Metropolitan governance and institutional cooperation

3.3.1 Joint development strategies

The development of joint development strategies for the Rabat-Sale-Temara conurbation has gradually become a necessity in the face of the challenges of urban sprawl, mobility and territorial cohesion. This development is taking place against a backdrop of growing awareness of the functional interdependencies between the various components of the metropolitan area.

The Rabat-Sale-Temara Urban Development Master Plan (SDAU), adopted in 2015, represents a major turning point in this concerted approach. This strategic planning document, drawn up under the aegis of the Agence Urbaine de Rabat-Sale, proposes an integrated vision of metropolitan development up to 2040 (Agence Urbaine de Rabat-Sale, 2015, p. 23). It emphasises shared objectives such as controlling urban sprawl, strengthening public transport and preserving natural periurban areas.

At the same time, structuring projects on a metropolitan scale have emerged, encouraging a concerted approach to development. The Bouregreg valley development project, piloted by the Bouregreg Valley Development Agency (AAVB), illustrates this dynamic. This ambitious project, which aims to create a new metropolitan area between Rabat and Sale, has necessitated the introduction of innovative coordination mechanisms between the various local authorities involved.

However, the effective implementation of these joint development strategies is still coming up against institutional and political obstacles. The difficulties associated with harmonising local urban planning documents (development plans) with the SDAU guidelines reflect the tensions between communal logic and metropolitan vision.

3.3.2 Supra-municipal coordination bodies

In response to the growing complexity of metropolitan issues, new supra-municipal coordination bodies have been set up, with the aim of reinforcing the coherence of public policies at the scale of the conurbation.

The Etablissement de Cooperation Intercommunale (ECI) Al Assima, set up in 2011, represents a significant step forward in this direction. Bringing together the municipalities of Rabat, Sale, Temara and several outlying municipalities, this body is responsible for coordinating transport, waste management and sanitation policies on a metropolitan scale (Moroccan Ministry of the Interior, 2021, p. 45).

The creation of the Rabat-Sale-Kenitra Development Agency in 2015 marks a new stage in metropolitan governance. This structure, which operates at the level of the administrative region, is responsible for promoting the region's economic development and attractiveness, by encouraging synergies between public and private players.

Despite these advances, challenges remain in terms of metropolitan governance:

a)   The persistent fragmentation of responsibilities between different levels of government (municipalities, prefectures, regions) and specialised agencies.

b)		Difficulties in coordinating sectoral strategies (transport, housing, development, etc.).

on a metropolitan scale.

c)		The lack of the supra-municipal coordination bodies' own financial resources, limiting their capacity for action.

To meet these challenges, discussions are currently underway on changes to the institutional framework for metropolitan governance.

Conclusion

This article on the functional integration of urban spaces in the context of the economic metropolis of Rabat-Sale-Temara has highlighted several significant dynamics and issues, while making theoretical and empirical contributions to the understanding of metropolisation processes in the context of emerging countries.

- Summary of the main results and theoretical contributions :

-Historical analysis of the development of the conurbation has revealed a complex process of spatial rapprochement and functional integration, marked by centrifugal dynamics and conurbation phenomena. These observations corroborate theories on the evolution of functional urban regions, while highlighting the specificity of urban trajectories in the Moroccan context.

-	The study of communication infrastructures and commuting patterns highlighted the intensification of flows and interdependencies between the different urban centres in the conurbation. These results confirm the importance of transport networks in structuring metropolitan areas.

-	Analysis of inter-sectoral economic complementarities has revealed the emergence of functional specialisations and productive synergies on a metropolitan scale. These observations enrich the theories on metropolitan productive systems, by highlighting the processes of economic integration in a context of transition.

-	Examination of the distribution of urban facilities and services with a metropolitan impact has highlighted the importance of the new commercial and cultural centres in the reconfiguration of spatial practices at the scale of the conurbation. These results help to refine our understanding of the dynamics of polarisation and multipolarisation in emerging metropolises.

- An analysis of concerted development strategies and supra-municipal coordination bodies has highlighted the advances and persistent challenges in metropolitan governance. These observations provide food for thought on the theoretical models of governance adapted to the complex realities of contemporary metropolises.

Research limitations and prospects :

Despite these significant contributions, the present study has certain limitations that open up prospects for future research:

- The limited availability of recent and reliable quantitative data on certain aspects of functional integration (intersectoral economic flows, fine-scale mobility) is a major constraint. The development of more effective data collection and analysis tools on a metropolitan scale appears to be a promising area for research.

- The study focused mainly on the dynamics within the conurbation, without examining relations with the surrounding periurban and rural areas. A more in-depth analysis of urban-rural interdependencies on the scale of the wider metropolitan region would enrich our understanding of the processes of functional integration.

- The comparative dimension with other Moroccan or North African metropolises remains limited in this study. In-depth comparative research would provide a clearer picture of the specificities and similarities of metropolitan dynamics in the region.

- The social and environmental impacts of functional integration processes have only been partially addressed. Future research could focus on the issues of socio-spatial inequalities and environmental sustainability associated with metropolisation.

- Finally, the analysis of innovations in metropolitan governance should be taken further, in particular by exploring the potential of participatory approaches and digital tools in the management of complex metropolitan areas.

In conclusion, this article sheds significant light on the dynamics of functional integration at work in the Rabat-Sale-Temara conurbation, thereby contributing to a richer theoretical and empirical understanding of the processes of metropolisation in the context of emerging countries. The research prospects identified pave the way for future investigations that will further refine our knowledge of the complex issues involved in the management and development of contemporary metropolitan areas.

Bibliography :

-        Agence Urbaine de Rabat-Sale (2015). Schema Directeur d'Amenagement Urbain de Rabat-Sale-Temara 2040. Rabat: AURS.

-        Rabat-Sale Urban Agency (2021). Enquete menages-deplacements de l'agglomeration Rabat-Sale-Temara. Rabat: AURS.

-        Agence Urbaine de Rabat-Sale (2022). Les pratiques de consommation dans l'agglomeration Rabat-Sale-Temara. Rabat: AURS.

-        Chapon, P.-M. (2020). Functional urban regions: concepts and applications. Economica.

-        Cusin, F., & Lefebvre, C. (2018). La question metropolitaine : les politiques a l'epreuve du changement d'echelle. PUF.

-        Direction de la Statistique du Maroc (2023). Annuaire Statistique de la Region Rabat-Sale-Kenitra. Rabat: HCP.

-        Dupuy, G., & Israelo, L. (2022). Les reseaux techniques urbains: nouvelles approches, nouveaux enjeux. Armand Colin.

-        EL AZIZI EL ALAOUI .A ,FATEH A. The Geomarketing and the competitiveness of the spaces urban Morocco, Revue Emirati Journal of Business, Economics and Social Studies, 2024, pp 66-78.

-        FATEH A. (2023) et Al. (Coordi.), Espace periurbains dynamiques et strategies de planification et de gestion coordination, Publication de la Faculte des Lettres et des Sciences Humaines, Universite Mohammed V Rabat, Rabat, 2023.

-        Abdelali FATEH (Coordi.) 2023: Transformations et formes d'adaptation des espaces ruraux, Tome 2, Publication de la Faculte des Lettres et des Sciences Humaines, Universite Mohammed V Rabat, 2 Tomes, Fr- Ar, 814p

-        Halbert, L. (2018). The metropolitan advantage. PUF.

- Halbert, L. (2021). Les metropoles : moteurs economiques a l'ere de la mondialisation. La Decouverte.

- Haut-Commissariat au Plan du Maroc (2022). Projections demographiques regionales 2020-2050. Rabat: HCP.

- Mohammed KAFSSI, Abdelali FATEH, Anas El ALAOUI Anas (2023). Planification stratégique durable et les enjeux de l'universite numerique au Maroc, Revue marocaine de l'education et de la recherche educative n° 10 /2023, 411383

- Lemarchand, N., & Mallet, S. (2019). Les temporalites urbaines a l'epreuve de la monopolisation. L'Harmattan.

- Moroccan Ministry of Higher Education (2022). L'enseignement superieur en chiffres 2021-2022. Rabat: MESRSFC.

- Moroccan Ministry of the Interior (2021). Report on inter-municipal cooperation in Morocco. Rabat: DGCL.

- Moroccan Ministry of Health (2023). Health map of the Rabat-Sale-Kenitra region. Rabat: MS.

- Pinson, G., & Demaziere, C. (2017). Integrated metropolitan systems. PUCA.

- Pumain, D., & Sanders, L. (2020). Theories and models of urbanisation. Economica.

- Sassen, S. (2021). The Global City: New York, London, Tokyo (3rd ed.). Princeton University Press.

- Chaimae TALHA-Abdelali FATEH, Mutations secretees par le processus de periurbanisation : cas du centre Oulad Tayeb, in Espaces periurbains : dynamique et strategies de planification, Coordination FATEH A. Rabat, 2023, 5-37.

_**The relationship between
sustainable strategic planning
and macro-spatial urban competitiveness
in Moroccan coastal cities**_

Anas EL AZIZI EL ALAOUI[II]

Resume :

Sustainable strategic planning is an essential approach for strengthening the macro-spatial urban competitiveness of Moroccan coastal cities. In a global context marked by growing environmental challenges, the adoption of holistic and proactive urban management is crucial. Given their strategic geographical position and considerable economic potential, Moroccan coastal cities face specific challenges related to their coastal location, such as coastal erosion, marine resource management and demographic pressure.

Sustainable strategic planning provides a conceptual and operational framework for integrating the principles of sustainable development into urban policies, reconciling the economic, social and environmental dimensions in a long-term perspective. This approach makes cities more resilient and competitive, making them more attractive to investment, stimulating innovation and offering a better quality of life for citizens.

Key words: Sustainable strategic planning, Urban competitiveness, Moroccan coastal cities, Sustainable development, Urban management, Urban resilience.

Introduction :

> **General and specific context :**

Sustainable strategic planning is emerging as an essential method for achieving urban competitiveness, particularly in Moroccan coastal cities facing complex challenges. This process aims to integrate the environmental, economic, social and urban dimensions in a holistic, long-term perspective. Its importance lies in its ability to enhance urban

---

II Faculty of Letters and Humanities, Mohammed V University, Rabat

competitiveness by improving the attractiveness, resilience and sustainability of cities. Key elements of this approach include internal and external analysis, visioning, target setting, strategy development, implementation and monitoring.

On the other hand, the macro-spatial urban competitiveness of Moroccan coastal cities is of crucial importance, encompassing economic, infrastructural, environmental and governance dimensions. However, these cities face specific challenges such as coastal erosion, marine resource management and demographic pressure, as well as opportunities for economic development and regional integration. Spatial planning therefore plays a key role in integrating sustainable strategic planning and urban competitiveness, through appropriate policies and tools, as well as effective governance involving the institutions concerned.

> **Issues and questions :**

Against a backdrop of urban globalisation and growing environmental challenges, *how can sustainable strategic planning contribute to strengthening the macro-spatial urban competitiveness of Moroccan cities, while ensuring balanced and sustainable territorial development?*

The central question can be divided into sub-questions:

*1.	What are the main issues and challenges facing Moroccan coastal cities in terms of macro-spatial urban competitiveness?*

*2.	How can sustainable strategic planning be put into practice in the specific context of Moroccan coastal cities, by integrating the environmental, economic, social and urban dimensions?*

*3.	What are the key success factors for the effective implementation of sustainable strategic planning to enhance the urban competitiveness of coastal cities (governance, stakeholder participation, funding, etc.)?*

*4.	What role can spatial planning policies and urban management tools play in supporting the integration of sustainable strategic planning and urban competitiveness?*

*5.	How can Moroccan coastal towns take advantage of their geographical position and economic potential, while preserving their* *natural environment and quality of life?*

6. *How can regional cooperation and integration into networks of cities contribute to strengthening the macro-spatial urban competitiveness of Moroccan cities?*

This problem and these research questions aim to explore the complex links between sustainable strategic planning, urban competitiveness and balanced territorial development in the specific context of Moroccan coastal cities.

I: Sustainable strategic planning as a method for achieving urban competitiveness

Sustainable strategic planning is emerging as an essential method for enhancing the competitiveness of cities in a globalised environment and in the face of the challenges of sustainable development. By integrating environmental, economic, social and urban dimensions in a holistic way, this approach enables cities to develop long-term strategies aimed at reconciling economic growth, the preservation of natural resources and the well-being of populations.

Sustainable strategic planning thus offers a conceptual and operational framework for improving the attractiveness, resilience and sustainability of urban environments, key factors in their competitiveness in a context of increasing competition between territories. Thanks to an integrated vision, the active participation of stakeholders and adaptation to local specificities, this methodology enables cities to position themselves advantageously in economic, environmental and social terms.

1.1   Concept of sustainable strategic planning and its dimensions

(environmental, economic, social, urban) Sustainable strategic planning is an integrated approach which aims to reconcile the imperatives of economic development, environmental protection and social equity in a long-term perspective.

Table (01): Dimensions of sustainable strategic planning

Dimension	Description	Key elements
Environmental	*Preservation of natural resources, sustainable management of*	▷ Protecting biodiversity ▷ Combating climate change ▷ Waste and water management

Dimension	Description	Key elements
	ecosystems, reduction of negative environmental impacts.	> Sustainable use of resources > Developing renewable energies
Economic	*Development of viable and competitive economic activities, fair distribution of wealth.*	> Creating sustainable jobs > Support for innovation and entrepreneurship > Promoting short circuits and the circular economy > Attracting responsible investment > Developing sustainable economic sectors
Dimension	**Description**	**Key elements**
Social	*Promoting community inclusion, cohesion and well-being, meeting basic needs.*	> Equitable access to essential services > Combating poverty and inequality > Promoting health and education > Respect for cultural diversity > Citizen participation and governance inclusive
Urban	*Planning and management of urban areas, integrating concerns relating to the built environment, transport, infrastructure and services.*	> Sustainable urban development > Sustainable mobility and public transport > Affordable, quality housing > Green spaces and meeting places > Risk-resistant infrastructure

Source : *Tanguay, G. A., Rajaonson, J., Lefebvre, J. F., & Lanoie, P. (2010). Measuring the sustainability of cities: An analysis of the use of local indicators. Ecological Indicators, 10(2), 407-418.*

This approach is based on four interdependent dimensions: ***The environmental dimension*** focuses on the preservation of natural resources, the sustainable management of ecosystems and the reduction of negative environmental impacts (Zuindeau, 2000).

The economic dimension focuses on the development of viable and competitive economic activities, while ensuring a fair distribution of wealth.

The social dimension aims to promote the inclusion, cohesion and well-being of communities by meeting their basic needs.

Finally, ***the urban dimension*** concerns the planning and management of urban areas, incorporating concerns about the built environment, transport, infrastructure and services (Thouvenot, 2007).

By integrating these four dimensions into sustainable strategic planning, we can adopt a systemic vision of urban development, promoting the resilience of cities in the face of environmental, economic and social challenges.

This approach encourages the active participation of stakeholders, cross-sectoral collaboration and the consideration of local specificities in the development of long-term strategies (Laigle, 2009). It thus offers a conceptual and operational framework for achieving a balance between the objectives of economic growth, environmental protection and social well-being in urban environments.

1.2 The importance of sustainable strategic planning in boosting urban competitiveness

Sustainable strategic planning is crucial to boosting urban competitiveness in a constantly changing world. This holistic approach aims to take account of economic, social and environmental issues in the development of cities, thereby ensuring their long-term sustainability and attractiveness. As Beal, Courville and Despres (2011, p. 23) point out, "*sustainable strategic planning is part of an integrated territorial development rationale, combining economic competitiveness, social equity and preservation of the natural and built environment*".

Indeed, urban competitiveness is no longer just about economics, but now also encompasses quality of life, residential attractiveness and environmental sustainability. A sustainable city that offers a healthy living environment, quality infrastructure and diversified employment opportunities will be better placed to attract and retain talent, business and investment.

In addition, sustainable strategic planning makes it possible to anticipate future challenges linked to climate change, resource scarcity and the energy transition. By incorporating these issues at the planning stage, cities can put in place adaptation and resilience measures, thereby guaranteeing their long-term competitiveness.

Finally, citizen participation and consultation with the various stakeholders are key elements of sustainable strategic planning (Beal et al., 2011, p. 37). This inclusive approach encourages public support for urban development projects, strengthens social cohesion and contributes to the city's attractiveness.

1.3 Key elements of sustainable strategic planning (internal/external analysis, vision, objectives, strategies, implementation, monitoring)

Sustainable strategic planning is a crucial process for ensuring the long-term sustainability and success of an organisation. This process involves in-depth analysis of the internal and external environments, definition of a clear vision, setting ambitious objectives, formulation of relevant strategies, effective implementation of these strategies and rigorous monitoring.

Figure (01): the sustainable strategic planning approach

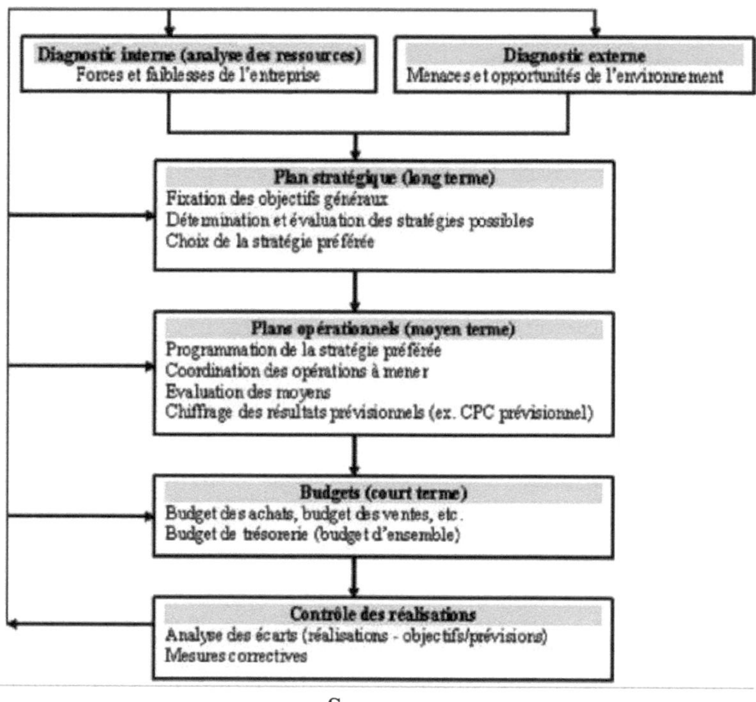

Source :
https://lewebpedagogique.com/oubejja/2020/03/26/planification-strategique/

Internal and external analysis is an essential step in identifying the strengths, weaknesses, opportunities and threats facing the organisation. The internal analysis assesses the organisation's resources, skills and capabilities, while the external analysis examines environmental factors such as market trends, competition, regulations and societal concerns (Leroy, 2008).

Figure (02): SWOT analysis of the external and internal environment

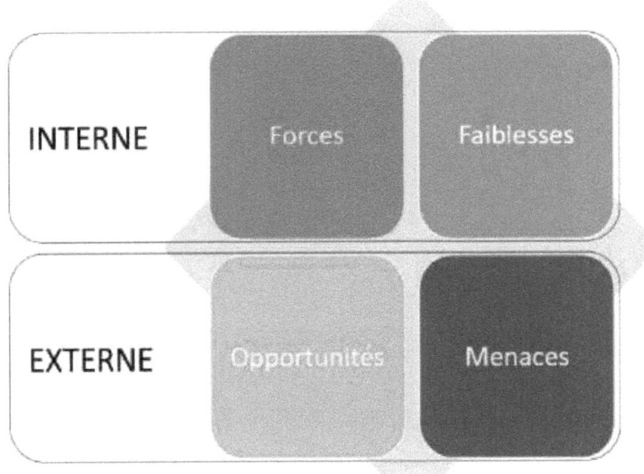

Source :

https://www.heflo.com/fr/blog/rh/comment-faire-une-analyse-swot/

The strategic vision is a key element that defines the desired future direction of the organisation and guides its decisions and actions. It must be inspiring, realistic and aligned with the organisation's values and sustainability principles.

Strategic objectives are specific, measurable, achievable, realistic and time-bound (SMART) targets that flow from the vision and reflect the organisation's priorities (Girerd-Potin, 1998). They must be balanced between economic, environmental and social dimensions to ensure a holistic approach to sustainable development.

Strategies are action plans designed to achieve set objectives. They must be adapted to the organisation's resources and capabilities, while taking account of the challenges and opportunities identified in the internal and external analysis.

Figure (03): Strategic urban planning process

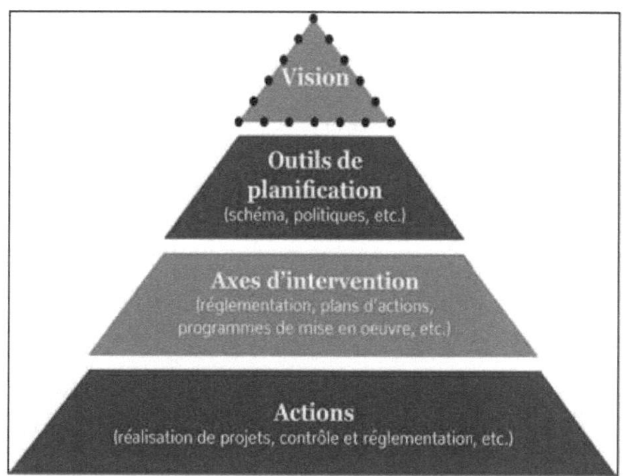

Source :
https://www.heflo.com/fr/blog/rh/comment-faire-une-analyse-swot/

Implementing strategies requires effective coordination of human, financial and material resources, as well as clear communication and involvement of all stakeholders. Appropriate management systems, efficient operational processes and an organisational culture conducive to change are essential for successful implementation.

Ongoing monitoring and evaluation are crucial for measuring progress, identifying deviations from objectives and adjusting strategies accordingly. Relevant performance indicators, both quantitative and qualitative, need to be defined to assess the achievement of economic, environmental and social objectives.

In short, effective sustainable strategic planning requires in-depth analysis, a clear vision, ambitious objectives, well thought-out strategies, rigorous implementation and constant monitoring. This will enable the organisation to adapt to emerging challenges, seize opportunities and make a positive contribution to sustainable development.

II: The macro-spatial urban competitiveness of Moroccan coastal cities

With over 3,500 kilometres of Atlantic and Mediterranean coastline, Morocco is home to a multitude of coastal towns that play a crucial role in the country's economic and social development. These cities are strategic gateways for international trade, foreign investment and tourism. However, in the current context of globalisation and increased competition between regions, the competitiveness of these coastal towns is of paramount importance in ensuring their attractiveness, sustainable growth and influence on a national and international scale. Macrospatial urban competitiveness encompasses a multitude of factors such as accessibility and connectivity, transport and communication infrastructure, quality of life, the business environment, skilled labour, economic poles of excellence, research and innovation centres, as well as governance and public policies. It is therefore essential to analyse these different dimensions in order to understand the strengths and challenges facing Moroccan coastal towns, and to identify the levers for action that will enhance their long-term competitiveness and attractiveness.

2.1 Concept of macro-spatial urban competitiveness and its dimensions (economic, infrastructure, natural environment/quality of life, governance)

The concept of macro-spatial urban competitiveness refers to the ability of a city or urban region to attract and retain economic activity, investment and talent, while offering a high quality living environment conducive to sustainable development. This competitiveness is based on several interdependent dimensions:

Economic dimension:

• *A diverse and dynamic economic fabric (business sectors, companies, jobs)*

• *Presence of centres of excellence and clusters of high added-value activities*

• *Access to regional, national and international markets*

• *Availability of skilled, productive labour*

• *An environment conducive to innovation and entrepreneurship*

Size of infrastructure:

• *Efficient transport networks (road, rail, air, sea)*

- *High-performance communications and information technology infrastructures*
- *Modern urban facilities (water, electricity, sewerage)*
- *Logistics infrastructures and multimodal platforms* **Dimension of the natural environment and living environment:**
- *Environmental quality (air, water, green spaces)*
- *Sustainable management of natural resources*
- *Urban amenities (public spaces, cultural and sports facilities)*
- *Diversified and affordable housing*
- *Safety and quality of public services*

Governance dimension :
- *Efficiency of urban management and public policies*
- *Collaboration between public, private and civil society players*
- *Ability to innovate and adapt to change*
- *Attractiveness and influence of the city on a national and international scale*
- *Taking sustainable development issues into account*

Macro-spatial urban competitiveness results from the synergy between these different dimensions, which enable a city to stand out, attract investment and talent, and promote the well-being of its citizens with a view to sustainable development.

2.2 Factors influencing the macro-spatial urban competitiveness of Moroccan coastal cities

Connectivity and accessibility are major assets for coastal cities, thanks to their port infrastructure, which facilitates trade and maritime transport. However, as Hamdouch, Khennas and Zaoual (2011) point out, the logistical performance of these cities is still hampered by deficits in road, rail and airport infrastructure, limiting their integration into regional and international transport networks.

The economic potential of Moroccan coastal towns is based mainly on activities linked to the sea, such as fishing, shipbuilding and beach tourism. However, the diversification of economic activities and the emergence of high value-added sectors, such as information and communication technologies, remain limited in these towns, hampering their competitiveness.

The natural environment and the quality of life also play a decisive role in the attractiveness of Moroccan coastal towns. However, many coastal towns face major environmental problems, such as pollution of coastal waters, degradation of marine and coastal ecosystems, and urban pressure on natural resources, all of which threaten their sustainability.

Urban governance is another key factor in the competitiveness of Moroccan coastal cities. These cities are faced with urban planning and management challenges, characterised by poor coordination between the different levels of governance, fragmented sectoral planning and limited participation of local stakeholders in decision-making processes.

2.3 Challenges and opportunities for boosting competitiveness

Morocco's coastal cities face a number of challenges in strengthening their macro-spatial urban competitiveness, but they also have promising opportunities to seize. The main challenges include :

The challenge of connectivity and accessibility, due to the lack of land, air and rail transport infrastructure linking these cities to the main national and international markets. This shortfall hampers the flow of trade and the mobility of people and goods.

The challenge of economic diversification, because many coastal towns are still heavily dependent on traditional activities linked to the sea, such as fishing and beach tourism, making their economies vulnerable to external shocks.

The environmental challenge, linked to urban pressure on natural coastal resources, marine pollution and the degradation of coastal ecosystems, threatening the sustainability of these cities and their attractiveness to tourists.

The challenge of urban governance, highlighted by the lack of coordination between the different levels of decision-making, fragmented sectoral planning and the limited involvement of local players in the development process.

Despite these challenges, Moroccan coastal towns also have

opportunities to boost their competitiveness, such as :

The opportunity to ***make the most of their strategic position*** as maritime gateways for international trade, by developing logistics platforms and integrated economic activity zones.

The opportunity to ***diversify their economies into emerging high-value-added sectors***, such as green industries, information and communication technologies, and business services.

The opportunity to ***preserve and enhance their natural and cultural heritage,*** by promoting sustainable tourism and developing coastal ecotourism projects.

The opportunity ***to strengthen their urban governance by adopting participatory approaches involving the various local players,*** promoting inter-municipal cooperation and integrating the principles of sustainable development into urban planning strategies.

To seize these opportunities and meet these challenges, a strategic vision and concerted mobilisation of public, private and civil society players are essential in order to strengthen the macro-spatial urban competitiveness of Moroccan coastal towns with a view to sustainable development.

III: The role of spatial planning in integrating sustainable strategic planning and urban competitiveness

Spatial planning plays a key role in the integration of sustainable strategic planning and urban competitiveness. These two key concepts cannot be effectively implemented without a comprehensive and coherent approach to the spatial organisation of human activities and the management of natural and heritage resources.

Spatial planning ensures a balanced and sustainable distribution of the population, economic activities, infrastructure and facilities throughout the country, taking account of regional and local specificities. This territorial approach, based on a long-term strategic vision and taking into account the economic, social and environmental dimensions, is essential for reconciling the competitiveness objectives of cities and regions with those of sustainable development (Mancebo, 2008). In this way, it helps to make the most of the assets and potential of each region, enhancing its attractiveness and influence, while preserving its natural resources and living environment for future generations.

3.1 The importance of spatial planning in sustainable urban planning

Spatial planning is of crucial importance in sustainable urban planning, as it enables the development challenges facing towns and regions to be addressed from an integrated, long-term perspective. Spatial planning provides a coherent framework for reconciling the economic, social and environmental objectives of sustainable development within a territorial approach.

One of the main contributions of spatial planning lies in its ability to organise and guide urban development in a rational and balanced way. As Merlin and Choay (2015) explain, this discipline makes it possible to guide the location and siting of economic activities, infrastructure, facilities and residential areas, taking account of natural resources, environmental constraints and local specificities. This helps to limit uncontrolled urban sprawl, preserve natural and agricultural areas, and encourage more sustainable use of land and energy resources.

In addition, spatial planning encourages an integrated and participatory approach to sustainable urban planning.

This discipline involves collaboration between the various local players, such as local authorities, businesses, associations and citizens, in order to define a shared vision of urban development and put in place strategies adapted to the realities on the ground.

Spatial planning also makes it possible to take into account the interdependencies and interactions between different territorial scales, from local to national, and even transnational. This multi-scalar approach is essential to ensure overall coherence in planning and sustainable development policies, by promoting complementarity and synergy between territories. Finally, spatial planning provides a framework for integrating sustainable strategic planning into urban management practices. This discipline enables the various sectoral strategies (economy, environment, mobility, housing, etc.) to be articulated within a coherent overall vision, by defining long-term objectives and putting in place monitoring and evaluation tools to adjust actions accordingly.

3.2 Policies and planning tools to support sustainable strategic planning and competitiveness

Spatial planning policies and tools play an essential role in supporting sustainable strategic planning and urban competitiveness. The main levers for action include :

Regional sustainable development plans (SRADDs), which define the major strategic guidelines for regional planning and development. These plans make it possible to integrate the principles of sustainable development into sectoral policies (transport, housing, environment, etc.) and to coordinate the actions of the various regional players around a shared vision.

Territorial Coherence Schemes (SCoT), which are strategic planning documents on an inter-municipal scale. SCoTs provide a framework for defining objectives for economic development, sustainable urban planning, the preservation of natural resources and the strengthening of social cohesion, while taking account of local specificities.

Local urban development plans (PLU), which set out the rules for land use and building layout at local level. These documents enable the strategic guidelines defined in the SCoTs to be implemented, by setting targets for urban densification, functional diversity, preservation of natural areas and quality of life.

Urban development projects (ZACs, eco-neighbourhoods, etc.), which are operational levers for putting sustainable development and competitiveness objectives into practice on the ground. These operations enable innovative projects to be developed in terms of energy efficiency, resource management and sustainable mobility, while at the same time making the most of the specific assets and potential of each area.

Local governance tools, such as development councils, citizen consultation and participation schemes, and public-private partnerships. These measures encourage a concerted and inclusive approach to defining and implementing sustainable development strategies, involving all local stakeholders.

The coherent articulation of these different spatial planning tools and policies, through a strategic vision and an integrated approach, will effectively support sustainable strategic planning and strengthen the competitiveness of towns and regions, while preserving natural resources and the living environment for future generations.

3.3   Role of institutions and urban governance in this integration

Institutions and urban governance play a key role in integrating sustainable strategic planning and competitiveness into spatial development policies. This integration requires effective coordination between the various players involved, a shared vision of territorial development and inclusive decision-making mechanisms.

Firstly, public institutions at different levels (national, regional, local) have a key role to play in defining strategic orientations and implementing planning policies. These institutions must ensure vertical coherence between the different levels of planning (national, regional, local) and horizontal coherence between the different sectoral policies (economy, environment, transport, etc.).

Secondly, urban governance, based on a participative and collaborative approach, is essential to encourage the support and involvement of all stakeholders (local authorities, businesses, associations, citizens) in sustainable development processes. This shared governance means that local issues and expectations can be taken into account, that the resources and skills of each stakeholder can be harnessed, and that the legitimacy and ownership of local projects can be strengthened.

Urban governance tools such as development councils, citizen consultation and participation schemes, and public-private partnerships play a crucial role in this integration. These mechanisms encourage a concerted and inclusive approach to defining and implementing sustainable development strategies, involving all local stakeholders.

In addition, collaboration between the various institutions and coordination of their actions are essential to ensure overall coherence in development policies. This coordination can take the form of territorial governance bodies (conferences, steering committees, etc.) or shared planning tools (territorial coherence schemes, territorial projects, etc.).

Finally, strengthening institutional capacities and improving urban governance are essential if the principles of sustainable strategic planning and competitiveness are to be effectively integrated into planning practices. This involves developing technical skills, setting up monitoring and evaluation systems, and promoting transparency and accountability in the management of territorial projects.

Conclusion:

Sustainable strategic planning is a key element in strengthening the macro-spatial urban competitiveness of Moroccan coastal cities. This holistic approach aims to harmonise economic development, preservation of the natural environment and social well-being, taking a long-term view. The judicious integration of sustainability principles into urban planning strategies makes it possible to maximise the geographical assets and natural resources of coastal towns, while minimising the negative impacts on fragile ecosystems.

Implementing sustainable strategic planning encourages the creation of resilient infrastructures, the optimisation of transport and urban mobility, the efficient management of natural resources and the promotion of sustainable economic activities. These elements contribute to improving the attractiveness of coastal cities to investors, talent and tourists, thereby strengthening their competitiveness on a macro-spatial scale.

In addition, the integration of social and environmental considerations into sustainable urban planning makes it possible to preserve the cultural and natural heritage of coastal regions, thereby promoting a better quality of life for residents and a unique experience for visitors. This holistic approach positions Moroccan coastal cities as attractive and sustainable destinations, giving them a significant competitive edge in the face of increasing global competition.

Bibliography :

- Aggeri, F., & Acquier, A. (2005). Convention theory as a tool for reflection on corporate social responsibility. Finance Controle

Strategie, 8(3), 97124.

-        Azadi, H., Ho, P., Hafni, E., Zarafshani, K., & Witlox, F. (2011). Multi-stakeholder involvement and urban green space performance. Journal of Environmental Planning and Management, 54(6), 785-811.

-        Badariotti, D., Benos, R., & Lajarge, R. (2018). Strategies d'amenagement durable desterritories   .   Presses Universitaires de Franche-Comte.

-        Beal, V., Courville, M., & Despres, C. (2011). Strategic and urban planning: Towards common knowledge? Presses de l'Universite Laval.

-        Bechet, B., Le Bissonnais, A., Ruas, A., Agator, J.-M., Andrieu, D., Aubry, S., ... & Schmitt, B. (2017). Measuring and valuing nature in the Schemas regionaux d'amenagement, de développement durable et d'egalite des territoires (SRADDET). Sciences Eaux & Territoires, (23), 28-33.

-        Berramed, D. (2002). Moroccan ports at the dawn of the 21st century. Publications de la Revue de Geographie du Maroc, Nouvelle Serie.

-        Bourdin, A. (2010). L'urbanisme d'apres crise. Editions de l'Aube.

-        Capron, M., & Quairel-Lanoizelee, F. (2007). Corporate Social Responsibility. La Decouverte.

-        Da Cunha, A. (2003). Developpement urbain durable, enjeux et pratiques. Editions Karthala.

-        Dauvergne, P., & Neville, K. J. (2009). The changing North-South and South-South political economy of biofuels. Third World Quarterly, 30(6), 1087-1102.

Education, heritage and architectural mix for sustainable urban development

Sustainable urban planning represents a major challenge for modern cities, which are called upon to reconcile economic development, environmental conservation and the well-being of their citizens. To achieve this ambitious objective, three fundamental pillars need to be carefully considered: education, heritage and architectural mix.

First and foremost, sustainable urban education is an essential lever for raising citizens' awareness of environmental challenges and training future generations in the trades and skills needed to build smart, resilient cities. With this in mind, the concept of the intelligent university is emerging as a promising model, aimed at radically rethinking teaching and research methods to meet the specific challenges of sustainable urban development in Morocco.

At the same time, preserving and promoting the city's rich architectural heritage represents a major challenge, but also a unique opportunity to combine the past and the future in a renewed vision of urban planning. The city of Rabat, with its historical heritage and architectural mix reflecting different cultural influences, offers a privileged field of study for exploring ways of reconciling the preservation of the treasures of the past with the development of contemporary architecture that respects the principles of sustainable development.

Sustainable urban education, an essential pillar for achieving intelligent citizenship in Morocco

Mohamed KAFSSI [3]

Resume :

Against a backdrop of increasing urbanisation and complex environmental, social and economic challenges, sustainable urban education is emerging as a crucial issue for Moroccan universities.

This educational approach aims to raise learners' awareness of the principles of sustainable development in urban areas and equip them to

---

[3] *Faculty of Education Sciences, Mohammed V University, Rabat*

contribute to the creation of more sustainable cities. This article explores the essential role of sustainable urban education in the realisation of the smart university model in Morocco. It examines the challenges facing Moroccan universities, the characteristics of the smart university, and the links between sustainable urban education and the sustainability, innovation and digital transformation pillars of this model.

The article also analyses the teaching strategies and practices that need to be implemented in order to effectively integrate sustainable urban education into Moroccan higher education. Finally, it highlights the expected impact of this approach on the creation of the intelligent university, in terms of environmental sustainability, innovation and competitiveness.

Key words: sustainable urban education, intelligent university, sustainable development, sustainability, higher education, Morocco, pedagogical innovation, digital transformation, sustainable city, 21st century skills, student training.

Introduction:

- *Context and issues :*

In a constantly changing world, marked by increasingly complex environmental, social and economic challenges, sustainable urban education is of crucial importance. In the face of rapid urbanisation, environmental degradation and growing social inequalities, it is imperative to educate responsible citizens who have a thorough understanding of sustainability issues and are capable of adopting environmentally-friendly lifestyles and practices.

Sustainable urban education aims to make learners aware of the challenges of sustainable development in urban areas, to instil in them values of respect for the environment and to give them the knowledge and skills they need to help create more sustainable and resilient cities. Moroccan universities face many challenges in a constantly changing environment. Rapidly evolving technologies, the emergence of new educational paradigms and society's growing expectations in terms of professional training and research pose major challenges.

Universities must adapt to the changing needs of the labour market, train students with cross-disciplinary skills and promote innovation and entrepreneurship. In addition, they are called upon to meet the challenges of sustainability, such as

The Group is committed to a number of key priorities, including the efficient management of resources, reducing its ecological footprint and raising awareness of environmental and social issues.

Faced with these challenges, the concept of the intelligent university is emerging as a promising response. The intelligent university is a model of higher education that holistically integrates the principles of sustainability, innovation and digital transformation. It aims to produce responsible students with critical thinking skills and an in-depth understanding of global issues, while promoting innovation and entrepreneurship. The Intelligent University focuses on the judicious use of information and communication technologies to improve teaching, research and resource management.

By adopting this model, Moroccan universities can meet the challenges of sustainability and innovation, while preparing their students to become key players in sustainable development in a constantly changing urban environment.

- ***Research objectives and questions :***

The main objective of this research is to highlight the essential role played by sustainable urban education in the implementation of the intelligent university model in Morocco. The aim is to demonstrate, through an in-depth analysis, that integrating the principles of sustainable urban education into higher education curricula is a fundamental pillar for achieving the transition towards more sustainable, innovative higher education establishments that are adapted to the current and future needs of Moroccan society.

To achieve the main objective, this study will address the following research questions:

a) *What are the main sustainability challenges facing Moroccan universities in a rapidly changing urban environment?*

b) *How can sustainable urban education help to meet these challenges and prepare students to become responsible citizens and key players in sustainable development?*

c)   What are the best practices and effective teaching strategies for integrating the principles of sustainable urban education into higher education curricula in Morocco?

d)   What are the potential obstacles to implementing sustainable urban education in Moroccan universities, and how can they be overcome?

e)   What impact is the adoption of sustainable urban education expected to have on the creation of the intelligent university, in terms of environmental sustainability, innovation and digital transformation?

I: Definitions of basic concepts and their theoretical underpinnings :

This section lays the theoretical foundations for the study by defining the key concepts of sustainable urban education and the smart university, and providing an in-depth review of the relevant literature. It begins by clarifying the principles, dimensions and approaches of sustainable urban education, highlighting its crucial role in raising awareness of sustainable development issues in urban areas. Next, the characteristics, components and models of the smart university are explored, highlighting its commitment to sustainability, innovation and digital transformation. Finally, the concepts of sustainable development and sustainability in the university context are examined, highlighting the environmental, social and economic dimensions that need to be integrated into higher education institutions.

1.1 Sustainable urban education: principles, dimensions and approaches

Sustainable urban education is a multidimensional concept that aims to raise people's awareness of the challenges of sustainable development in urban areas and equip them to contribute to the creation of sustainable cities. According to Diemer and Mulnet (2017, p. 23), this form of education is based on three fundamental principles: the integration of environmental, social and economic dimensions, the promotion of critical thinking and problem-solving, and the encouragement of civic engagement and collective action.

In environmental terms, sustainable urban education focuses on understanding the challenges of managing natural resources, pollution, biodiversity and climate change in urban environments (Lazzeri and Baurieau, 2021, p. 47). It aims to raise learners' awareness of the impact of their lifestyles on the environment and to instil in them more sustainable behaviours, such as reducing waste, using energy efficiently and preserving urban green spaces.

The social dimension of sustainable urban education encompasses aspects such as social equity, environmental justice, community cohesion and citizen participation (Rabie, 2016, p. 118). It aims to develop an understanding of the social and environmental inequalities present in urban areas, as well as the skills needed to promote social inclusion and the active participation of citizens in decision-making about their environment.

In economic terms, sustainable urban education focuses on understanding sustainable production and consumption models, as well as promoting the circular economy and green entrepreneurship. It encourages learners to think about the economic impacts of their consumption choices and to explore more sustainable alternatives.

Several pedagogical approaches are used in sustainable urban education, such as project-based learning, experiential learning and community service learning. These approaches encourage the active involvement of learners, the acquisition of practical skills and an understanding of local issues related to sustainable development in urban areas.

1.2 Intelligent university: features, components and models

The concept of the intelligent university, also known as the sustainable university or the green university, refers to a model of higher education that integrates the principles of sustainability, technological innovation and digital transformation into its educational, research and management practices. A smart university is characterised by its commitment to sustainable development, its ability to adapt to environmental, social and economic changes, and its strategic use of information and communication technologies (ICT) to improve learning and resource management.

The key components of an intelligent university can be grouped into three main areas: environmental sustainability, digital transformation and pedagogical innovation. In environmental terms, an intelligent university aims to reduce its ecological footprint by implementing sustainable practices such as energy efficiency, waste management, biodiversity conservation and the promotion of sustainable modes of transport.

Digital transformation is an essential pillar of the intelligent university, which involves integrating ICT into all aspects of teaching, research and administration. This can mean the use of e-learning platforms, virtual classrooms, digital libraries and intelligent data management systems.

Pedagogical innovation is also at the heart of the intelligent university model, with the adoption of new student-centred teaching and learning approaches such as project-based learning, experiential learning and collaborative learning. These approaches aim to develop cross-disciplinary competences that are essential for the 21st century, such as critical thinking, problem-solving, creativity and the ability to work in a team.

Several intelligent university models have been proposed, each focusing on specific aspects. The Green University model focuses primarily on environmental sustainability, while the Digital University model places greater emphasis on digital transformation and the use of ICT. Other models, such as the Sustainable University, take a more holistic approach, integrating the environmental, social and economic dimensions of sustainability.

Whatever the specific approach adopted, the common objective of these models is to transform universities into sustainable, innovative institutions that are adapted to the current and future challenges facing society, while preparing students to become responsible citizens and key agents of change.

1.3 Sustainable development and sustainability in the university context

The concept of sustainable development has become increasingly important in the academic context over the last few decades. Initially defined by the Brundtland Report in 1987 as *"development that meets the needs of the present without compromising the ability of future generations to meet their own needs"* (World Commission on Environment and Development, 1987, p. 51), sustainable development is based on three interdependent pillars: the environment, society and the economy.

In the university context, environmental sustainability refers to efforts to reduce the ecological footprint of campuses, promote the efficient use of natural resources and raise awareness of environmental issues among the university community. This can take the form of initiatives such as the construction of eco-energetic buildings, the introduction of waste management systems, the promotion of sustainable transport and the preservation of biodiversity on campus.

The social dimension of sustainability in universities concerns the equity, inclusion and well-being of all members of the university community, as well as a commitment to social responsibility. This involves putting in place policies and programmes to promote diversity, combat discrimination, facilitate access to higher education and encourage community engagement.

In economic terms, sustainability in universities refers to the responsible management of financial resources, the promotion of sustainable procurement practices and the training of students in the challenges of the green economy and sustainable entrepreneurship. Universities are encouraged to adopt more sustainable business models, invest in sustainable development research projects and train their students in the skills needed to succeed in a more environmentally friendly economy.

Integrating the principles of sustainable development into universities requires a holistic and systemic approach, involving all aspects of university life, from teaching and research to operations and governance. Universities are called upon to play a leading role in promoting sustainable development, by training future leaders and contributing to the creation of knowledge and innovative solutions to the environmental, social and economic challenges facing the world.

II: Links between sustainable urban education and the intelligent university

This section explores the close links between sustainable urban education and the realisation of the intelligent university model. It highlights the essential contribution of sustainable urban education to the three pillars of the intelligent university: environmental sustainability, digital transformation and pedagogical innovation. The synergies between the sustainability principles promoted by sustainable urban education and the objectives pursued by the intelligent university are highlighted. Finally, this section discusses the importance of integrating the environmental, social and economic dimensions of sustainability into all aspects of higher education in order to fully realise these synergies.

2.1. Contributions of sustainable urban education to the pillars of the intelligent university

Sustainable urban education plays a fundamental role in achieving the three pillars of the intelligent university: environmental sustainability, digital transformation and educational innovation.

By raising students' awareness of the challenges of sustainable development in urban environments, sustainable urban education helps to train responsible citizens who are aware of their impact on the environment and capable of adopting more sustainable behaviour and lifestyles.

This environmental awareness is essential to achieving the sustainability objectives of the Intelligent University, such as reducing the ecological footprint, managing resources efficiently and promoting sustainable modes of transport.

In addition, by encouraging the use of information and communication technologies (ICT) in experiential learning and project-based learning, sustainable urban education supports digital transformation, a central pillar of the smart university. Students acquire essential digital skills, while exploring innovative solutions to urban challenges using emerging technologies.

Finally, sustainable urban education fosters pedagogical innovation by encouraging the adoption of student-centred approaches to teaching and learning, such as project-based learning, experiential learning and collaborative learning.

These innovative teaching methods enable students to develop crucial cross-disciplinary skills for the 21st century, such as critical thinking, problem-solving, creativity and the ability to work as part of a team, while tackling concrete issues linked to sustainable development in urban areas.

2.2. Synergies between the principles of sustainability and the objectives of the intelligent university

There is a close synergy between the sustainability principles promoted by sustainable urban education and the objectives pursued by the intelligent university model. First of all, environmental sustainability, which is an essential pillar of sustainable urban education, is perfectly in line with the intelligent university's efforts to reduce its ecological footprint and promote environmentally-friendly practices.

In addition, the social dimension of sustainability, with its emphasis on equity, inclusion and civic participation, ties in with the Intelligent University's objectives in terms of social responsibility and community involvement. By training students to be aware of social and environmental issues, the intelligent university is helping to create a fairer, more sustainable society.

Finally, the economic aspect of sustainability, which promotes sustainable production and consumption models and the circular economy, is aligned with the Smart University's efforts to train students in the skills required in a green and sustainable economy. Sustainable entrepreneurship and innovation are also key themes in both areas.

2.3. Integrating environmental, social and economic dimensions into higher education

To fully realise the synergies between sustainable urban education and the intelligent university, it is essential to integrate the environmental, social and economic dimensions of sustainability into all aspects of higher education. This means incorporating these issues into the curricula, research activities, operations and governance of universities. In terms of teaching, it is crucial to include courses and modules dedicated to sustainable development, sustainable urban education and environmental, social and economic issues in urban areas. This content can be integrated transversally into different disciplines, or offered in the form of programmes.

In the field of research, universities should encourage and support interdisciplinary projects on issues related to urban sustainability, sustainable innovation and environmentally-friendly technological solutions. Collaboration between researchers from different disciplines is essential if these complex issues are to be tackled holistically.

Finally, university operations and governance must reflect the principles of sustainability, implementing sustainable practices in campus, infrastructure and resource management. This can include initiatives such as energy efficiency, waste management, the promotion of sustainable transport and the adoption of responsible procurement policies.

By fully integrating the environmental, social and economic dimensions of sustainability into teaching, research and operations, Moroccan universities will be able to make a significant contribution to the realisation of the intelligent university model and train students with the skills and knowledge needed to meet the challenges of sustainable development in urban areas.

III: Implementing sustainable urban education in Moroccan universities

This section focuses on the strategies and concrete actions needed to implement sustainable urban education in Moroccan universities. It examines the regulatory framework and national policies promoting sustainability in Moroccan higher education, providing a context conducive to this transition. Next, effective pedagogical strategies and practices, such as project-based learning, experiential learning and community service learning, are proposed to integrate the principles of sustainable urban education. Finally, the essential role of all stakeholders, including students, teachers, the administration and the local community, in achieving this transition is emphasised.

3.1. Regulatory framework and national policies promoting sustainability in higher education

Morocco has made significant efforts to integrate the principles of sustainable development into its national policies, including in the field of higher education. The National Strategy for Sustainable Development 2015-2020 (Ministry of Energy, Mines, Water and the Environment, 2014) stresses the importance of education for sustainable development and encourages higher education institutions to play a key role in training the necessary skills.

In addition, the Strategie Nationale de l'Environnement et du Developpement Durable 2022-2030 (Ministere de la Transition Energetique et du Developpement Durable, 2022) emphasises the need to strengthen education and awareness of the environment and sustainable development at all levels, including higher education. This strategy encourages universities to integrate environmental and sustainability issues into their curricula and research activities.

3.2. Teaching strategies and practices for integrating sustainable urban education

To effectively implement sustainable urban education in Moroccan universities, it is essential to adopt appropriate teaching strategies and practices. Project-based learning, experiential learning and community service learning are particularly relevant approaches for addressing sustainable development issues in urban areas.

Project-based learning encourages students to work in teams on concrete projects related to urban challenges, such as the design of sustainable solutions for transport, waste management or sustainable urban planning. This approach enables students to develop practical skills while gaining an in-depth understanding of local issues.

Experiential learning involves learning activities in a natural environment, such as field visits, case studies or fieldwork in urban neighbourhoods. This method enables students to come face-to-face with the realities on the ground and understand the challenges of sustainable development in urban environments.

Finally, community service learning encourages students to become actively involved in sustainable development projects in their local communities. By working closely with local stakeholders, students gain a practical understanding of the issues and help to find practical solutions.

3.3. Role of stakeholders (students, teachers, administration, local community) in the transition to an intelligent university

The transition to the intelligent university model requires the involvement and collaboration of all stakeholders, including students, teachers, the university administration and the local community.

Students play a central role in this transition, by adopting sustainable behaviours and lifestyles, taking an active part in sustainable development initiatives on campus and getting involved in sustainable urban education projects. Their commitment and motivation are essential to the success of these initiatives.

Teachers have a responsibility to integrate the principles of sustainable urban education into their courses and teaching methods. They should be trained in innovative pedagogical approaches, such as project-based learning and experiential learning, and encourage students to develop critical thinking and transversal skills. The university administration plays a key role in implementing policies and programmes that promote sustainability on campus, such as waste management, energy efficiency and the promotion of sustainable transport. It must also support research initiatives related to sustainable development and encourage interdisciplinary collaboration.

Finally, the local community, including local authorities, non-governmental organisations and businesses, must be involved in the transition to the smart university. Partnerships with these actors can provide practical learning opportunities for students, promote knowledge transfer and contribute to the development of sustainable solutions adapted to local needs. By putting in place a favourable regulatory framework, adopting innovative teaching strategies and practices, and involving all stakeholders, Moroccan universities will be able to rise to the challenge of integrating sustainable urban education and thus contribute to the realisation of the smart university model.

IV: Impacts and benefits of sustainable urban education for the intelligent university

This final section highlights the significant impacts and benefits of integrating sustainable urban education into the realisation of the smart university model. The environmental benefits, such as the reduction of the ecological footprint and the sustainable management of resources, are explored. Social benefits, such as raising awareness of responsible citizenship and developing cross-disciplinary skills, are also highlighted. Finally, the positive economic implications, such as increased innovation, competitiveness and employability, are discussed, demonstrating the holistic benefits of this approach for Moroccan universities.

4.1. Environmental benefits: reduced ecological footprint, sustainable resource management

Integrating sustainable urban education into Moroccan universities will have a significant environmental impact, helping to achieve the sustainability objectives of the intelligent university. By raising students' awareness of the challenges of sustainable development in urban areas, this educational approach will encourage them to adopt more environmentally-friendly behaviour and lifestyles.

Students trained in the principles of sustainable urban education will be better equipped to help reduce the ecological footprint of university campuses, by adopting practices such as reducing waste, using energy efficiently, preserving green spaces and promoting sustainable transport.

In addition, these students will be able to implement sustainable resource management strategies, encouraging the use of recycled materials, the reuse of resources and the introduction of rainwater recovery or composting systems. These practices will help to reduce the environmental impact of the universities and promote more responsible use of natural resources.

4.2. Social benefits: raising awareness of responsible citizenship, developing cross-disciplinary skills Sustainable urban education will also have significant social benefits as part of the transition to the intelligent university. By tackling the issues of social equity, environmental justice and civic participation, this educational approach will raise students' awareness of their role as responsible citizens and agents of change.

Students will gain an in-depth understanding of the social and environmental challenges facing urban communities, as well as the skills needed to actively engage in the search for sustainable solutions. They will be encouraged to take part in sustainable development projects within their local community, thereby strengthening their sense of social responsibility and civic engagement.

In addition, the innovative teaching methods used in sustainable urban education, such as project-based learning and experiential learning, will enable students to develop cross-disciplinary skills that are essential for the 21st century. These include critical thinking, complex problem solving, creativity, the ability to work in teams and communication and leadership skills.

4.3. Economic implications: innovation, competitiveness and employability

Integrating sustainable urban education into Moroccan universities will also have positive economic implications, helping to boost innovation, competitiveness and student employability.

By training students who are aware of the challenges of sustainable development in urban areas, Moroccan universities will be preparing a skilled workforce tailored to the emerging needs of the green economy and the circular economy. These students will be able to propose innovative solutions to environmental and social challenges, thereby contributing to the competitiveness of Moroccan businesses and organisations.

In addition, the emphasis placed on sustainable entrepreneurship and innovation as part of sustainable urban education will encourage students to develop innovative ideas and projects, thereby promoting the creation of sustainable businesses and economic growth. These entrepreneurial initiatives can contribute to the creation of green jobs and the transition to a more sustainable economy in Morocco.

Finally, the cross-disciplinary skills acquired by students through sustainable urban education, such as critical thinking, complex problem solving and the ability to work in a team, will enhance their employability on the labour market. These skills are highly valued by employers and will contribute to students' competitiveness in a constantly changing professional environment.

By incorporating sustainable urban education into their curricula, Moroccan universities will be able to train a skilled workforce tailored to the emerging needs of the sustainable economy, while encouraging innovation, entrepreneurship and competitiveness, thereby helping to achieve the objectives of the intelligent university.

Conclusion:

This study explored the crucial role of sustainable urban education in the realisation of the smart university model in Morocco. The results showed that integrating the principles of sustainability, pedagogical innovation and digital transformation inherent in sustainable urban education is essential to meeting the challenges facing Moroccan universities in a constantly changing urban environment. By raising students' awareness of the challenges of sustainable development in urban areas and equipping them with cross-disciplinary skills, sustainable urban education contributes directly to the achievement of the pillars of the intelligent university, namely environmental sustainability, innovation and digital transformation.

Although this study has provided an in-depth analysis of the links between sustainable urban education and the smart university, certain limitations need to be highlighted. Firstly, the research focused mainly on the theoretical and conceptual context, and practical case studies would be needed to assess the effectiveness of the proposed strategies in different Moroccan university contexts. Secondly, the study focused on Morocco, but it would be interesting to explore the transferability of these results to other countries and regional contexts.

Beyond that, a number of recommendations can be made to policy-makers, university administrators and teachers in the field of higher education in Morocco:

• To develop a regulatory framework and national policies to encourage the integration of sustainable urban education into the curricula and research activities of Moroccan universities.

• To encourage and financially support initiatives aimed at training teachers in innovative pedagogical approaches linked to sustainable urban education, such as project-based learning and experiential learning.

• Promote interdisciplinary collaboration and partnerships with local stakeholders (municipal authorities, NGOs, businesses) to provide practical learning opportunities for students and develop sustainable solutions tailored to local needs.

• Set up sustainability awareness and training programmes for all members of the university community, including students, teachers and administrative staff.

• Adopting sustainable practices in the management of university campuses, such as energy efficiency, waste management and the promotion of sustainable transport, to serve as a concrete example for students.

Bibliography :

• Amaral, L. P., Martins, N., & Gouveia, J. B. (2015). Quest for a sustainable university: a review. International Journal of Sustainability in Higher Education, 16(2), 155-172.

• Amaral, L. P., & Martins, N. (2015). Cooperation between sustainable higher education: global trends. International journal of sustainable higher education, 16(1), 105-123.

• Becker, S., Cummins, M., Davis, A., Freeman, A., Hall, C. G., & Ananthanarayanan, V. (2018). Interdisciplinary education and training for sustainable development: transdisciplinary pedagogy of experiential learning. Discourse and Communication for Sustainable Education, 9(1), 21-35.

- Bricas, N., & Denechere, F. (2020). Vers des systemes alimentaires durables: trajectoires et enjeux critiques. Editions Que.
- World Commission on Environment and Development. (1987). Our Common Future. Oxford University Press.
- Diemer, A., & Mulnet, D. (2017). L'education au developpement durable dans l'enseignement superieur : enjeux, defis et perspectives. Presses Universitaires de France.
- Leal Filho, W., Frankenberger, F., Salvia, A. L., & Azeiteiro, U. (2019). Education for sustainable development in higher education: global trends. Peter Lang University Press.
- Leal Filho, W., Pace, P., & Manolas, E. (2017). Student engagement in sustainability education in higher education. Springer.
- Leal Filho, W., Skanavis, C., Kounbuntan, D., Brandli, L. L., Shiel, C., Neves, F. O., ... & Pace, P. (2018). Teaching and learning about education for sustainable development in higher education. Springer.
- Ministry of Energy, Mines, Water and the Environment. (2014). Strategie Nationale de Developpement Durable 2015-2020. Government of Morocco.
- Ministry of Energy Transition and Sustainable Development. (2022). Strategie Nationale de l'Environnement et du Developpement Durable 2022-2030. Government of Morocco.

The architectural mix and the challenge of sustainable heritage, the case of the city of Rabat

Malak JABLI [4]

Resume :

The architectural mix of Rabat, the capital of Morocco, presents a fascinating case study in the context of sustainable heritage and urban evolution. The city has a diverse architectural heritage, encompassing structures ranging from the historic medina and Art Deco colonial

---

[4] *Faculte des Lettres et des Sciences Humaines, Universite Mohammed V Rabat.* [2] The French protectorate in Morocco is the protectorate exercised by France in the Cherifian Empire, i.e. present-day Morocco. It was established by the Treaty of Fes, on 30 March 1912, between the Third French Republic and Sultan Moulay Abdelhafid and ended on 18 November 1955 with the return of Sultan Mohammed Ben Youssef, the future King of Morocco, Mohamed V. The protectorate came to an end on 02 March 1956.

quarters to post-independence modernist architecture and contemporary buildings.

This juxtaposition of styles and epochs raises crucial questions about the preservation and enhancement of the built heritage from a sustainable development perspective. The challenges are manifold: not only do we need to conserve historic buildings while adapting them to modern functional requirements, but we also need to integrate new buildings harmoniously into the existing urban fabric, while promoting ecologically responsible and culturally relevant architecture.

This issue is part of a wider context of maintaining urban identity and developing sustainable cultural tourism. Local and national authorities are thus faced with the need to implement balanced conservation and rehabilitation policies, particularly in sensitive areas such as the medina, while at the same time developing urban development projects that reconcile modernity with respect for heritage.

The stakes are high: the aim is to transform this architectural mix into an asset for sustainable urban development, capable of meeting contemporary needs while preserving Rabat's unique historical and cultural heritage.

Keywords : Architectural mix; Sustainable development; Sustainable heritage; Morocco; Rabat.

Introduction :

Moroccan cities are extremely diverse, given that Morocco is a country in the Arab world and the wider Maghreb, situated in north-west Africa and bordered to the south by Mauritania, to the north by the Straits of Gibraltar, to the west by the Atlantic Ocean and to the east by Algeria, to the West by the Atlantic Ocean and to the East by Algeria. Because of its strategic geographical position and its location at the meeting point of three worlds that it links, Africa, Europe and the Mediterranean, it is considered to be a true crossroads country.

In addition, thanks to its thousand-year history, rich in events and the many civilisations that have crossed its path, Morocco is rich in many different cultures, beliefs, languages, architecture and religions, particularly following the French colonisation of the country from 1912[2].

Indeed, the country undeniably reveals a multitude of cultures and the mix of architectural styles, i.e. the architectural mix of the country, is based on its past but also and above all, on its own cultural identity.

The city of Rabat, Morocco's capital, on the shores of the Atlantic Ocean, is an administrative city, but above all it is a cultural city steeped in history, where the jewels of the past stand side by side with the most modern, environmentally-friendly projects. Rabat's architectural, urban and landscape heritage is rich and diverse, and safeguarding and promoting it has been a priority for several decades.

"The city of Rabat has always made the preservation and enhancement of its heritage, inherited from previous generations, a priority shared by the various stakeholders in the economic and social life of the city" *(Rabat, capitale moderne et ville historique : un patrimoine en partage - Plan gestion - Proposition sur la liste de patrimoine mondial soumis par le Royaume Maroc, January 2011, p. 04).* (Rabat, capitale moderne et ville historique : un patrimoine en partage - Plan de gestion - Proposition d'inscription sur la liste du patrimoine mondial soumis par le Royaume du Maroc, Janvier 2011, p. 04)

Certainly, from the dawn of his new reign, His Majesty King Mohammed VI[5] , may God assist him, has given impetus to a powerful and intense drive to enhance and promote heritage throughout the Kingdom, but most particularly in the country's capital, thanks in particular to the creation of the Foundation for the Safeguarding of Rabat's Cultural Heritage in 2012. The Royal desire and vision was, and still is, to make Rabat a city that is mindful of and attentive to the balance between culture and nature, as well as development and heritage.

However, what are the real foundations and pillars of the architectural mix and rich heritage of the Moroccan capital? What are the issues involved in preserving its heritage? How can sustainable development be reconciled with heritage conservation and enhancement? What are the strategies, tools and principles of sustainable heritage, particularly in the city of Rabat?

---

[5] The current King of Morocco, who succeeded his late father Hassan II in 1999.

The aim of this study is to highlight the architectural mix in Morocco, particularly in the city of Rabat, and to shed light on the issues involved in preserving heritage, while addressing the latter as one of the pillars of sustainable development. The aim of this article is to clarify the relationship between sustainable development and heritage, by understanding the term sustainable heritage and illustrating it in terms of the Moroccan capital.

Finally, this article is divided into three main parts. The first focuses on the city of Rabat as an imperial city with a rich history and architectural heritage, highlighting its urban development and different architectural styles. This section also clarifies the role of heritage conservation in the capital, as well as identifying any social and economic problems associated with the conservation of this heritage.

The second part aims to clarify the concept of sustainable development and the impact of construction and buildings on the environment, as well as the importance of combining sustainable development and heritage, and therefore the benefits of sustainable heritage.

The final part analyses and evaluates Moroccan regulations and the urban planning measures taken to safeguard Rabat's heritage in the long term, as well as the structures created to improve its management.

In short, this section aims to understand the prospects for development from a heritage point of view and the future visions for sustainable heritage in Rabat.

1. Rabat, a city with a rich architectural heritage.

For countless and diverse reasons, the city of Rabat is destined to play a decisive and important role in Morocco's economic development, and must increasingly assert its role as a growth pole, both at regional and national level. So, in order to play the role assigned to it and to achieve these objectives, Rabat needs to meet the ever-increasing needs imposed by Morocco's demographic growth and urban explosion, but also to face up to a number of challenges, particularly those of territorial attractiveness, economic competitiveness and social cohesion, while respecting its environment.

1-1 Urban development in Rabat.

Rabat, an imperial city par excellence, has been developing since the beginning of the XXeme century and has been marked by several key dates and different urban planning documents that have marked the city's urban history. Among the main documents that have marked the development of Rabat are, of course, the Prost plan[6] of 1914 and the Ecochard plan[7] of 1955, both of which were drawn up during the French protectorate and led to the creation of the new town by extending the old town, i.e. the medina. Subsequently, two master plans for the urban development of Rabat-Sale were produced, in 1971-1972 and 1995, which more or less brought the urban explosion in the capital under control, as in all other cities in the country and the world.

Morocco's capital has remained confined to the status of an administrative city for a very long time, but it has always had the ambition of becoming the kingdom's international city of knowledge and culture, obviously under the impetus of His Majesty King Mohammed VI, who wanted to turn the city of his Alaouite ancestors into a truly international cultural capital, and who therefore launched the "Rabat, City of Light, Moroccan Capital of Culture" programme in 2014, turning it into a veritable open-air development site. This programme is based primarily on a participatory approach aimed at preserving and promoting the cultural and civilisational heritage of the city of Rabat, while protecting its environment, improving access to services and infrastructure, consolidating good governance and, finally, preserving and upgrading the city's urban fabric. Supervised by the Societe Rabat Region Amenagement[8] , this programme is structured

---

[6] Henri Prost was an architect and town planner who was born in France on 25 February 1874 and died on 17 July 1959. Following the establishment of the French protectorate in 1912, he drew up the development plan for the city of Rabat in 1914 under the orders of Marechal Hubert Lyautey, the French politician who governed colonial Morocco (Archiwebture, 2024).

[7] Michel Ecochard was a French architect and town planner who was born on 11 March 1905 and died on 24 May 1985. He was director of the town planning department in Morocco, then under the French protectorate, and produced a number of master plans for Moroccan towns, including that of Rabat-Sale in 1995 (Archiwebture, 2024).

[8] More commonly known as RRA, Rabat Region Amenagement is a company set up in November 2014 to manage the development programme for the city of Rabat from 2014 to 2018. It is the public operator responsible for the operational implementation of the "Rabat, City of Light, Moroccan Capital of Culture" programme. (Rabat Region Amenagement, n.d.).

around seven different objectives, each corresponding to a development area, in order to encourage and ensure urban and social development within the capital, as shown below:

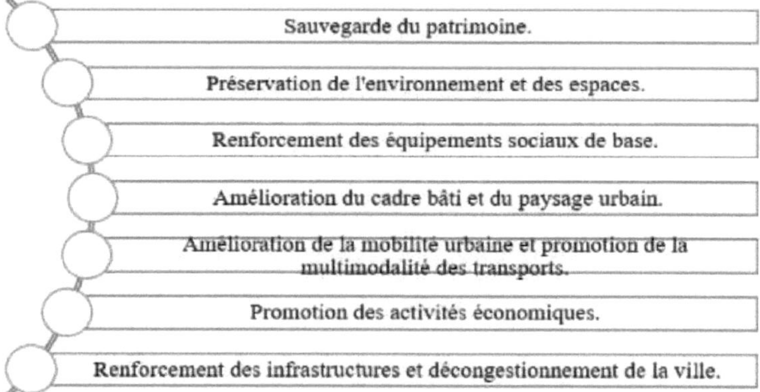

Figure -1- : The main thrusts of the "Rabat, City of Light, Moroccan Capital of Culture" programme
Source: Reproduced from (Rabat Region Amenagement, n.d.)

Without a doubt, Rabat has continued to develop over the years and is still experiencing unprecedented urban growth, thanks in particular to major new projects such as the Grand Theatre de Rabat, the Mohammed VI Tower, the Rabat-Sale airport modernisation and extension project, the new Ibn Sina hospital, the Mohammed VI Polytechnique University, the new bus station and the Cité des Metiers et des Compétences. Of course, Rabat's major urban project was conceived and designed without neglecting its cultural heritage, which ultimately earned it the title of "Rabat, African Capital of Culture" in 2022, making it the first city and capital on the continent to be honoured with this distinction.

1-2 Rabat's architectural mix.

The architecture of Rabat's buildings reflects the city's history, marked by the various civilisations that have shaped its architectural style. In fact, Rabat's architecture is undeniably varied and diverse, and the following architectural styles can be distinguished:

Table -1 : Different architectural styles in Rabat

The Art Nouveau style.	The Neo-Traditional style.	The Neo-classic style.	The ArtDeco style.
Born at the end of the XIXeme century, this style is characterised by a diversity of materials and animal and plant forms as sources of inspiration. Built during the French protectorate, the buildings in Rabat in this style are distinguished by their decorative elements and their curves, but remain present in the city with only two buildings in this style.	Born towards the end of the 19eme century, this style was born of a revoluti on the political, architectural and urban policies pursued by France in North Africa, and accompanied the arri val of urban planning. It is characterised by the use of simple, sober volumes. enhanced by decorative elements borrowed from local architecture. It is this style that is mainly	Used between 1750 and 1850, this style is characterised by an arrangement of fagades giving a strong visual unity to the buildings. buildings, with symmetry and different floors. Moreover This style can also be found on boulevard Mohammed V in the capital Moroccan.	Appearing around 1910, this style is much more sober than Art Nouveau and is recognised by the organisation of its fagades (alignment, re petition and symmetry) and verticality. The decoration uses repeated geometric motifs: chevrons, triangles, squares, etc. In addition, a large number of buil dings in
	is used in the Diour Jamaa district.		Flaps are made in this style after the First World War.
Modern style.	**The Brutalist style.**	**The Post-Modern movement.**	**Official architecture.**
Called switching The Modern	Primarily based on the use of untreated	First appeared in the early 70s as a	An entreneo-like blend classical in the

movement, this style expresses the primacy of function over form. Simple and unadorned, it nevertheless introduces the terrace roofs, banded windows and cornices. Unlike Art-Deco, the emphasis is on horizontal lines and white. Rabat, the The first buildings in this style were constructed from 1929 onwards.	materials, beton brut de This style is based on a return to expressive architecture linked to the social context and the local area. Rabat, the concrete facades include referencesa traditional architecture with the use of zelliges for example.	goes beyond the modern trend or complements it question some of its postulates These include collage, eclecticism, juxtaposition, humour and exaggeration. Two buildings from this movement were built in Rabat in 1980.	organisation of the facades and neo-traditional in its symmetry, this is the offi cial style of the architecture of the French protectorate, with a certain attachment to one of the This is the best characteristic of Arab architecture, i.e. exterior sobriety (clean lines and simplicity of contours and fagades).

Source: Reproduced from (Plan d'amenagement et sauvegarde du patrimoine de la ville de Rabat, Rapport justificatif, 2017, pp. 23-38).

As can be seen from the table above, one of Rabat's distinguishing features is undoubtedly its architectural mix and the different architectural styles found within its boundaries.

1-3 Heritage preservation issues.

Whether it's an opportunity to pass on our culture and identity, a chance to create jobs and business, or a way to protect our planet and its ecosystems, preserving our heritage can be a successful strategy for territorial development and strategic planning. In fact, preserving and enhancing the heritage aspect of towns and cities has a number of advantages, such as, for example, according to the Fondation du Patrimoine (Portail du patrimoine, 2023) :

- Heritage is a way of passing on history and culture: a testimony to the many different architectural styles present in towns and cities, heritage reflects the history, culture and identity of different civilisations and events, past or present.

- Heritage creates opportunities for people to share and come together: it is often an opportunity and a place for people to share and meet, and an opportunity to participate in the good life together and the development of links between citizens.

- Heritage as a lever for development and attractiveness: helps to create tourist appeal, given that the heritage on offer remains an important factor in the choice of destination and therefore makes a major contribution to the development of tourism in the area.

- Heritage creates jobs: promoting and talking about heritage is an opportunity to maintain or create jobs indirectly, by contributing to the attractiveness of tourism through the development of the tourist sector (restaurants, cafés, etc.).

- Heritage plays a part in preserving the environment: heritage is not just limited to historic monuments, but also to natural sites and landscapes. Preserving our heritage is a positive way of contributing to the ecological transition and helping to preserve the environment and biodiversity.

"Heritage protection is therefore part of a responsible and sustainable approach, as it aims to reuse and restore rather than destroy or build elsewhere. (Heritage Portal, 2023)

In the case of the city of Rabat, the preservation of its heritage has enabled the Rabat-Sale-Kenitra region to become Morocco's 6eme tourist destination, behind Marrakech, Agadir, Casablanca, Fes and Tangiers, thus demonstrating its tourism potential. Rabat accounts for 80% of the region's tourism offering, with a total of 602,360 tourists in 2012, compared with just 16,290 in Sale, its neighbouring city, according to the Ministry of Tourism and Crafts. Moreover, a total of 257,472 overnight stays were recorded in tourist accommodation establishments in Rabat during the first four months of 2024, compared with 243,653 between January and April 2023, an increase of 6%, according to the Tourism Observatory. This performance can be explained, in particular, by the city's development and the cultural and heritage attractions it offers. Thanks to its close proximity to the public, heritage is a key lever in cultural policy, revealing its full potential. Enhancing and protecting our heritage means helping to pass it on from generation to generation, and to preserve our collective memory.

1-4 The disadvantages of heritage preservation.

The preservation and enhancement of heritage, like any territorial policy, can have negative effects and some disadvantages, for example:

- There is no doubt that talking about heritage, and maintaining and improving it, comes at a huge cost and requires a great deal of expenditure. Obviously, the amounts awarded and allocated to heritage are inequitable and unequal, depending on the city and region. Moreover, sometimes these restoration, conservation or improvement projects have no return, profitability or direct impact on the living conditions of the local inhabitants.

- The policy of heritage enhancement does not necessarily achieve its intended economic objectives, given its incompatibility with government policies. Often, national, regional or local interests may clash, and a heritage enhancement project may conflict with a national development project and policy, with the result that no state support (subsidies, expertise, logistical resources, etc.) is forthcoming.

- The conversation and enhancement of heritage can lead to certain inequalities and create a kind of exclusion in the distribution of earnings and income. Indeed, the benefits generated by heritage improve the living conditions of people in the tourist sector without improving those of low-income residents.

- The political aspect of heritage enhancement can, in turn, have negative effects on economic development, given that the choice of heritage assets to be promoted and enhanced can sometimes be the subject of economic and political conflicts, depending on the preferences of decision-makers, which can be diametrically and radically opposed.

The conversation, preservation and enhancement of heritage does not necessarily contribute to economic development, but requires an analysis of the potential, an impact study and an in-depth knowledge of the economic and social dynamics of the area. Of course, the attitude and perception of the population towards heritage is crucial to its economic enhancement, and detailed knowledge of its assets is a significant and obvious factor in economic development.

2. Heritage and the challenges of sustainable development.

A heritage approach cannot be dissociated from sustainable development. Indeed, as we briefly emphasised in the first part of our study, heritage conservation helps to preserve the environment and biodiversity. However, what does the concept of sustainable development really mean? And what is the link between sustainable development and heritage?

2-1 Sustainable development.

Towards the end of the 1980s, the concept of sustainable development was defined by a famous phrase coined by the then Prime Minister of Norway, Harlem Gro Brundtland:

"Development that meets the needs of the present without compromising the ability of future generations to meet their own needs. (Gro Brundtland, H., 1987).

This definition highlights the need for, and importance of, a twofold solidarity: intergenerational solidarity over time, for future interests, as well as solidarity in the present.

In addition, the European Commission's 1992 Maastricht Treaty gives a further definition of the concept of sustainable development, describing it as a policy aimed at ensuring the continuity of social and economic development over time, while respecting the environment and, of course, without compromising or damaging natural resources:

"Determined to promote the economic and social progress of their peoples, within the framework of the completion of the internal market and the strengthening of cohesion and environmental protection, and to implement policies ensuring parallel progress in economic integration and in other areas". (Extract from the European Commission's Treaty of Maastricht, 7eme , 1992).

On the other hand, sustainable development cannot be put in place and established in people's minds unless a number of factors are taken into account, including education, the exchange of experience, training and culture, all of which can bring about changes in behaviour and lifestyles that will shake up people's habits.

The concept of sustainable development is closely related to the notion of development, which is based on the notion of progress. The concept is therefore defined as a development process based on economic efficiency, respect for the environment and, above all, social progress. It combines and reconciles three major, often antagonistic but necessary foundations: protection of the environment goes hand in hand with human development, and economic development must not deplete human and natural resources.

Figure -2- : The pillars of sustainable development

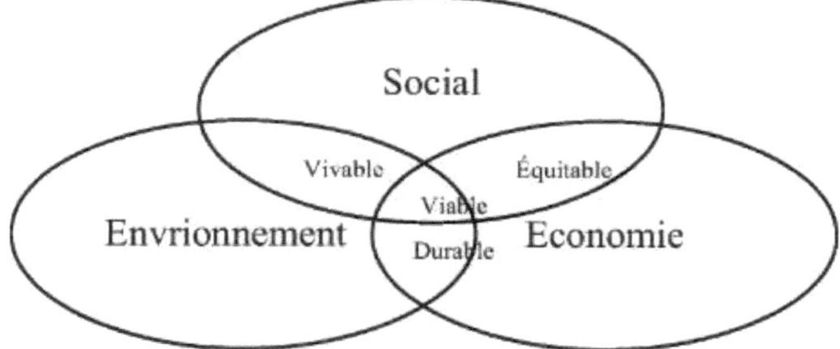

Source: Reproduced from (Laboudi, S., 2013)

There is no doubt that sustainable development is a comprehensive approach with a vision for the future, capable of linking the short term with the long term. In other words, it involves broadening the temporal and spatial horizon, taking future generations into account. This concept has various aims and objectives, including, for example: social equity and solidarity, respect for ecosystems, preservation of biodiversity, access to knowledge, responsible production and consumption, economic efficiency and, most importantly for our purposes, protection of our heritage.

In fact, from 1995 onwards, when discussions focused solely on the environmental aspect, they gradually turned to the urban environment, and it was at the first biennial city conference in Lyon, which brought together a number of urban planners, that the question of the need for a link between sustainable development and cities was raised, leading to an approach based on three key and important principles:

Table -2- : Approach to the biennial cities event in Lyon

Ecosystem approach.	A heritage approach.	A participatory approach.
Ma in the city as a consumer of natural resources.	Principle giving priority to cultural access in the interests of safeguarding cultural diversity, heritage features and improving the living environment.	A principle that gives priority to the political aspect and aims to take greater account of local realities.

Source: Reproduced from (Laboudi, S., 2013)

There is a clear link between heritage and sustainable development, as the latter helps to highlight culture and improve the management of architectural, landscape and urban assets.

2-2 Impacts of buildings on the environment.

Buildings (houses, blocks of flats and private homes) consume an enormous amount of energy, and the construction sector undoubtedly has an impact on the environment and the climate. In reality, the sector generates 23% of air pollution, 40% of drinking water pollution and 50% of waste. It is also a huge user of resources, most of which are non-renewable.

In fact, the repercussions affect biodiversity and have a significant and damaging long-term effect on flora and fauna. In terms of air pollution, buildings are responsible for almost 39% of carbon dioxide emissions[9] . As far as waste is concerned, a massive amount is produced annually by the construction sector in Morocco, as elsewhere in the world, with an estimated 41.9 million tonnes of building and public works waste per year, i.e. 90% of the total waste produced, with no recovery in

---

[9] Carbon dioxide, more commonly known by its molecular formula CO2, plays an active role in the respiration of living organisms and the photosynthesis of plants. Over the past 150 years, the amount of CO2 in the atmosphere has increased tenfold. From 180 ppm in 1850, this has risen to almost 400 ppm in 2016. Although an exhaustive list of the effects of the increase in carbon dioxide in the atmosphere has yet to be drawn up, certain consequences can already be seen (GEO magazine, 2022).

Morocco, ending up in public landfill sites because, at the same time, recycling is not compulsory.

Fortunately, the construction sector is changing, albeit slowly, but increasingly incorporating the environmental approach and aspect. Indeed, green buildings[10] have become a trend supported by government funding, and many companies are turning to modern technologies such as solar energy. Energy-efficient renovation of existing buildings can also significantly reduce energy consumption and greenhouse gas emissions .[11]

On the other hand, it is undeniable that traditional Moroccan architecture is flourishing thanks to the preservation of ancestral building techniques that enable buildings to be passively cooled, and also to the use of local materials that guarantee sustainable, environmentally-friendly construction.

2-3 Principles and tools for sustainable heritage.

The protection of outstanding heritage assets (great natural landscapes or historic monuments) is seen as an intrinsic contribution to the well-being of humanity. Heritage certainly makes an important contribution to sustainable development in its various dimensions: reducing poverty and inequality, improving health and safety services or preserving natural resources. Often, heritage is also a key asset for economic development, particularly through tourism.

"While it is common to associate the protection of "culture" with "sustainable development" - the idea of preserving a heritage for the benefit of present and future generations is indeed inherent to both concepts - the classic definition of sustainable development and the content of this notion as specified by certain founding texts generally overlook the importance of culture for the development of societies."

---

[10] Green buildings are those which, in their design, construction and operation, reduce or eliminate negative impacts on the environment. They are also capable of having a positive impact on our climate and natural environment. Such environmentally oriented characteristics do not, however, prevent buildings from serving their purpose and functioning properly. (Energy Management, 2021).

[11] Certain gases naturally present in the earth's atmosphere help to retain heat close to the earth's surface. They are known as "greenhouse gases" (GHGs) and consist mainly of water vapour, carbon dioxide, methane, nitrous oxide and ozone. Without these gases, the average temperature on Earth would be -18°C and life on Earth would be impossible (Ministere de l'Environnement, de la Lutte contre les changements climatiques, de la Faune et des Parcs, Quebec, 2024).

(Guevremont, V., 2014, p. 161).

It is undeniable that sustainable development and heritage are extremely closely linked, and the assets that make up the latter are considered to be durable assets that aim to satisfy the needs of both present and future generations, which is why they are considered to be sustainable heritage. This is why a multitude of programmes and actions have been implemented, such as the inclusion of historic monuments on the national heritage list, the classification of sites of interest and the signing of development agreements for several urban areas, as part of a global process.

3. Sustainable heritage as a development perspective.

As mentioned above, at the dawn of his new reign, King Mohammed VI wanted to make Rabat a modern capital and a city of history, thanks to a shared heritage. This approach has been reinforced by the city's inclusion on UNESCO's World Heritage List[12] since July 2012, thus affirming the genuine and indisputable universal value of the kingdom's capital and highlighting its renowned development.

3-1 Urban planning and regulatory measures to protect Rabat's heritage.

Integrity, authenticity and requirements in terms of management and protection, several measures have been drawn up to protect historic monuments, urban ensembles, archaeological sites and urban landscapes. As far as Rabat is concerned, it is considered to be *"a place of considerable interchange of influences"* and *"an outstanding example of an architectural ensemble illustrating a significant period in human history"*, according to UNESCO criteria[11] II and IV, which has led to the creation of various laws, measures and regulations for the city. Among the laws that have enabled and continue to enable heritage protection are three successive laws:

---

[12] The United Nations Educational, Scientific and Cultural Organization (UNESCO) is an international institution created in 1945 in the aftermath of the Second World War. It is a specialised agency of the United Nations Organisation (UNO) and its purpose is to maintain security and peace through science, culture and education, by promoting collaboration among the nations in order to further universal respect for the rule of law, justice, fundamental freedoms and human rights (UNESCO: World Heritage Convention, 2024).

- The Dahir of 13 February 1914: relating to the conservation of historic monuments, works of art and antiquities in the Cherifian Empire and inscriptions, and the protection of areas surrounding monuments and sites.

- The Dahir of 21 July 1945: relating to the conservation of historic monuments and sites, works of art and antiques and the protection of old towns and regional architecture.

- Dahir n° 1-80-341, promulgated by law n° 22-80: relating to the conservation of historic monuments and sites, inscriptions and works of art and antiquities.

Furthermore, in order to meet these commitments, a Heritage Development and Conservation Plan (P.A.S.P) was launched in 2017, covering a discontinuous area made up of two components of the city of Rabat, both of which are on UNESCO's World Heritage List: the new town, comprising[13] an area of 146 hectares, which is part of the Rabat Hassan district, and the Habous Quarter located in Diour Jamaa, covering an area of 4 hectares, which is part of the Agdal Ryad district. This study is considered to be an ambitious document, driven by a strong desire to prevent the disappearance or irreversible damage to the historic centre of Rabat.

"They point out that 'economic challenges, urban renewal and property speculation are exerting strong pressure on the fabric of the 20th century' and state that 'it is above all a question of supporting the future of the city, in order to ensure continuity with the past and to make a lasting mark, in space and in the collective memory, on the city of tomorrow'" (Plan d'amenagement et sauvegarde du patrimoine de la ville de Rabat, Rapport justificatif, 2017).

The PASP is therefore first and foremost a planning and management tool that makes it possible to articulate all the actions to be taken to safeguard the architectural, urban and landscape heritage and the measures to be taken to implement them.

Subsequently, in November 2021, a Special Development Plan (SDP)

---

[13] In order to be included on UNESCO's World Heritage List, proposed sites must be of outstanding and universal value and meet at least two of the ten criteria (UNESCO: World Heritage Convention, 2024).

was launched, which led to the legal protection of the area and improved and more thorough decision-making. As a result of the PAS, any project (restoration, renovation, repair, rehabilitation or new build) must undergo a heritage impact assessment (EIP) or an assessment by the Historic Monuments and Sites Inspectorate[14] , before it can be approved and authorised for implementation. And during the construction phase, unannounced visits by an inspection committee may be carried out to ensure that the work is compliant and that the sites in question remain in a state of protection.

Urban planning and regulatory measures, as well as the actions undertaken by the city of Rabat, have made it possible to safeguard, protect and enhance its heritage. Actions undertaken in the field of governance and culture, such as the establishment of safeguard plans for the medina, the city centre and Diour Jamaa, urban communication campaigns and information campaigns, have also helped to highlight this rich heritage.

3-2 Development prospects and future visions.

It is clear and indisputable that safeguarding our heritage is essential, and that despite the many challenges facing cities and the pressures they face (economic, political and social), protecting culture must be an even greater priority. It is therefore necessary to continue with the steps, actions and measures undertaken to date, in order to ensure appropriate management over the long term.

As we have seen, this initiative has boosted tourism in the city of Rabat, with a significant increase in overnight stays in tourist accommodation establishments, as shown below:

Table -3-: Change in overnight stays in establishments of tourist accommodation in Rabat from 2014 to 2023.

2014	2015	2016	2017	2018	2019	2020	2021	2022	2023
639 272	618 951	629 685	714 700	778 629	785 260	339 080	371 724	669 137	819 827

Source: Reproduced from (Royaume du Maroc, Ministere du Tourisme, de l'Artisanat et de l'Economie Sociale et Solidaire, 2024).

---

[14] It is one of the representatives of the Ministry of Culture that carries out the inspection of historic sites and monuments and therefore manages the national heritage. (Istasse, M., 2013, pp. 377-384).

There is no doubt that the number of overnight stays at tourist accommodation establishments in Rabat has risen sharply, with the exception of 2020 and 2021, due to the Covid-19 health crisis. These results are extremely encouraging for the city, particularly the 819,827 overnight stays achieved in 2023, and the ambition is to reach the million tourist mark by the end of 2024, in order to achieve sustainable development in the Moroccan capital and improve living conditions for its inhabitants. Moreover, according to a former local councillor and member of parliament, the responsibility does not lie solely with the local authorities and the State, but is shared with the citizens, who have a duty to play their part in enhancing and preserving the city's heritage. In his words: *"The problem is not just administrative, it is also and above all human. Everyone needs to get involved in the project to enhance the city's urban heritage. And that's why I'm proposing the creation of an association dedicated to preserving this heritage"*. (Enhancing Rabat's urban heritage, 2011).

Indeed, it is also important to involve citizens and to include a participatory approach in the heritage process, but, of course, to attach particular importance to research and studies with the aim of better understanding and safeguarding the heritage.

Conclusion

This study reminds us of the importance of heritage in the sustainable development of cities, particularly the city of Rabat, the capital of Morocco. Preserving and safeguarding heritage makes a major contribution to the transmission of heritage and the common good from one generation to the next, as well as to the revitalisation of the urban unit. This is why it is vital to reconcile heritage and sustainable development, something that Rabat was quick to grasp and which has enabled it to become one of the most visited cities in the kingdom, as well as one of the most developed. Thanks to a strong political will and a number of measures and actions put in place, Rabat is a city that promotes sustainable heritage, with an extremely rich past and a highly promising future.

Bibliography

- *Kingdom of Morocco, Ministry of Culture (January 2011). Rabat, modern capital and historic city: a shared heritage - Management Plan - Nomination to the World Heritage List submitted by the Kingdom of Morocco.*

- *https://fr.wikipedia.org/wiki/Henri_Prost. (s.d.).*

- *Archiwebture. (2024, 05 16). Retrieved from Centre d'archives d'architecture contemporaine: https://archiwebture.citedelarchitecture.fr/archive/fonds/F RAPN02_PROST/inventaire/objet-21564*

Archiwebture. (2024, 05 16). Retrieved from Centre d'archives d'architecturecontemporaine : https://archiwebture.citedelarchitecture.fr/archive/fonds/F RAPN02_ECOMI

Rabat Region Amenagement. (s.d.). Retrieved from https://www.rra.ma Ministere de l'amenagement du territoire national, de l'Urbanisme, de l'Habitat et de la Politique de la Ville; Wilaya de la region de Rabat-Sale-Kenitra; Prefecture de Rabat; Commune urbaine de Rabat; Agence urbaine de Rabat-Sale (2017). Plan d'amenagement et sauvegarde du patrimoine de la ville de Rabat, Rapport justificatif. Rabat. Portail du patrimoine. (2023). Retrieved from Fondation du patrimoine: https://www.portailpatrimoine.fr/qui-sommes- us

Gro Brundtland, H. (1987). Our Common Future by the World Commission on Environment and Development.

Extract from the European Commission's 7th Maastricht recital. (1992).

Laboudi, S. (2013). Towards sustainable urban renewal: the case of the Yacoub Al Mansour neighbourhood, Rabat.

Guevremont, V. (2014). Sustainable development in the service of cultural heritage: a propos de la Convention pour la Sauvegarde du Patrimoine Culturel Immateriel. Ethnologies, 36 (1-2).

UNESCO: World Heritage Convention (2024).

Retrieved from https://whc.unesco.org/fr/criteres/

UNESCO: World Heritage Convention (2024).

Retrieved from https://whc.unesco.org/fr

Kingdom of Morocco, Ministere du Tourisme, de l'Artisanat et de l'Economie Sociale et Solidaire. (2024). Retrieved from Tourist arrivals : Record attendance: https://mtaess.gov.ma/fr/tourisme/chiffres-cles/

GEO magazine. (2022, 10 28). Retrieved from https://www.geo.fr/environment/co2-what-is-carbon-dioxide-193560#

Energy Management. (2021). Retrieved from KNX: https://sustainabilityknx.org/fr/articles-fr/what-is-a-green- building-and-what-are-its-features/

Ministere de I'Environnement, de la Lutte contre les changements climatiques, de la Faune et des Parcs, Quebec (2024). Recupere sur

https://www.environnement.gouv.qc.ca/air/questce-ges.htm#:~:text=They%20are%20appeles%20"%20gas%20 a,'ozone°%20(O3).

Istasse, M. (2013). Politique patrimoniale a Fes et rdle de l'Inspection des monuments historiques. Centre Jacques- Berque.

Enhancing Rabat's urban heritage (2011). Retrieved from https://lematin.ma/journal/2011/Initiative_Valoriser- le-patrimoine--urbain-de-Rabat/157120.html

Part III

Socio-spatial equality and urban intelligence

The advent of the sustainable development era has profoundly shaken up the paradigms of urban and regional planning. Over and above purely environmental considerations, this new holistic approach aims to radically rethink the way in which urban areas are designed and managed, placing socio-spatial equity and territorial intelligence at the heart of the concerns.

In the Arab world, where galloping urbanisation has often been accompanied by a worrying fragmentation of the urban fabric, the question of spatial justice between central districts and peripheral areas is particularly acute. It is now essential to analyse in depth the segregationist dynamics at work and to identify the levers of action that can be used to reduce these socio-spatial divides, which are a source of inequality and tension.

With this in mind, the concept of urban intelligence is emerging as a particularly promising tool for rethinking the governance of cities and optimising the distribution of resources and services across the territory. The case study of Casablanca, a megalopolis in the throes of change, will provide a concrete understanding of the key role that information and communication technologies can play in reducing socio-spatial disparities and building fairer, more inclusive cities.

Sustainable urban planning in the Arab world: the current situation and the challenge of spatial justice between the centre and the periphery

Rachida BAILI[1] Jamila OUARDI [15 16]

Resume :

This article examines the current state and future prospects of sustainable urban planning in the Arab world, focusing on the issue of spatial justice between urban centres and peripheries. The study begins by tracing the evolution of the concept of sustainable urbanism in the specific context of Arab countries, analysing its theoretical foundations and practical applications. It then looks at the centre-periphery dichotomy characteristic of many Arab cities, exploring the resulting socio-spatial disparities and their implications for sustainable urban development.

The article identifies and analyses the main challenges facing Arab cities in implementing sustainable urban planning policies, including institutional, economic and cultural obstacles.

Finally, it proposes an integrated model of sustainable urban planning adapted to the specificities of the Arab world, aimed at promoting more equitable and environmentally friendly urban development. This model is based on concrete examples and case studies from various Arab metropolises.

Tags: sustainable urban planning, Arab world, spatial justice, centre-periphery dichotomy, equitable urban development, urban planning, sustainable cities, urban policy, socio-spatial disparities, urban environment

Introduction: Context and issues of sustainable urban planning in the Arab world

Sustainable urban planning in the Arab world is a complex, multidimensional paradigm that requires a holistic approach that takes

---

1- _Doctor in Geo-environment Member of Laboratoire Geo-Ingenierie et environnement (LGIE)Faculte des sciences Meknes,_
[16] - _Teacher-researcher, CRMEF, Casablanca Settat._

account of the region's socio-cultural, economic and environmental specificities. In a context of rapid and often anarchic urban growth, characterised by strong demographic pressure and major environmental challenges such as the scarcity of water resources and the degradation of ecosystems, Arab cities are faced with the imperative of rethinking their urban development models.

The issue of sustainable urban planning in this region revolves around the need to reconcile the demands of modernisation and economic competitiveness with the principles of environmental sustainability and social equity, while preserving the cultural and heritage identity of the urban fabric. This quest for balance comes up against structural obstacles such as fragmented urban governance, pronounced socio-spatial inequalities between centres and peripheries, and the prevalence of unsustainable consumption patterns.

The crucial issue of spatial justice, which underpins the equitable distribution of resources, services and opportunities within the urban space, is emerging as a fundamental area for reflection and action for sustainable urban planning in the Arab world.

This problem raises fundamental questions about the ways in which the concepts and practices of sustainable urban planning, developed mainly in Western contexts, can be adapted and put into practice in the specific realities of Arab cities, which are marked by unique socio-spatial dynamics, governance structures and environmental challenges.

The aim of exploring these issues is to contribute to the development of innovative urban strategies and policies, capable of promoting sustainable and inclusive urban development that meets the aspirations of local populations while taking a global view of sustainability.

I. The evolution of sustainable urban planning in the Arab context: theoretical and practical foundations

I.1 Origins and development of the concept of sustainable urban planning

The concept of sustainable urban planning has its roots in the environmental and social movements of the twentieth century, but its formalisation and integration into urban policies crystallised around the

Brundtland report of 1987.

This report, entitled "Our Common Future", laid the foundations for a global reflection on sustainable development, including its urban dimension (Gro Harlem Brundtland, 1987, p. 43). Sustainable urban planning then developed as a response to the environmental, social and economic challenges posed by galloping urbanisation and unsustainable urban development models.

The work of Francois Ascher (2001, p. 79) on 'sustainable urban development' has helped to theorise this approach, emphasising the importance of integrated urban planning that takes account of the complex interactions between urban systems and their environment.

The Aalborg Charter (1994), followed by the United Nations conferences on human settlements (Habitat II in 1996 and Habitat III in 2016), marked crucial stages in the institutionalisation of sustainable urban planning at international level, by defining frameworks for action and concrete objectives for cities (Catherine Charlot-Valdieu and Philippe Outrequin, 2009, p. 22).

I.2 Adapting sustainability principles to Arab urban realities

Adapting the principles of sustainable urban planning to the Arab context requires a critical reappraisal of Western models and their adjustment to local specificities. The work of Youssef Diab and Salem Dahir (2011, p. 156) highlights the particular challenges facing Arab cities, including water management in arid environments, preserving urban heritage in the face of modernisation, and reconciling traditional urban structures with the demands of contemporary life.

The study by Nabil El-Khoury and Pierre-Antoine Barthel (2017, p. 89) on eco-neighbourhoods in the Arab world highlights the need for a contextual approach, incorporating vernacular knowledge of bioclimatic architecture and resource management.

This adaptation also involves redefining urban governance, as suggested by the research of Mona Fawaz (2019, p. 203), who argues for greater citizen participation and decentralisation of decision-making processes in urban planning, in order to better respond to the specific needs of local communities.

I.3 Comparative analysis of theoretical and practical approaches in different Arab countries

A comparative analysis of sustainable urban planning approaches in the Arab world reveals a diversity of strategies and levels of implementation. Eric Verdeil's work (2018, p. 112) on sustainable urban policies in the Maghreb and the Mashreq highlights significant disparities between the Gulf countries, which focus on high-tech, energy-intensive urban projects, and the Maghreb countries, which favour approaches that are more rooted in local socio-economic realities.

The study by Safaa Monqid and Aziz Iraki (2020, p. 67) on Morocco and Egypt highlights the tensions between national ambitions for urban sustainability and local constraints, particularly in terms of institutional capacity and financial resources.

Pierre-Arnaud Barthel and Valerie Clerc's research (2015, p. 178) on sustainable new towns in the Arab world reveals the limits of top-down approaches and the importance of co-constructing urban projects with local stakeholders. These comparative analyses highlight the need for a flexible and adaptive approach to sustainable urban planning, capable of taking account of the diversity of Arab urban contexts while adhering to a common vision of sustainability.

II. The centre-periphery dichotomy in Arab cities: analysis of socio-spatial disparities

II.1 Characterisation of urban centres and peripheries in Arab cities

The morphology of contemporary Arab cities is characterised by a marked dichotomy between urban centres and peripheries, the result of a complex historical evolution and specific socio-economic dynamics.

The work of Jean-Francois Troin (2006, p. 219) on the metropolises of the Maghreb highlights the coexistence of densely populated and often degraded historic centres (medinas), colonial town centres with predominantly administrative and commercial functions, and rapidly expanding and often unplanned suburbs.

This spatial configuration can also be seen in Mashreq cities, as highlighted by Eric Denis' study (2007, p. 78) on Cairo, where the distinction between a multifunctional centre and residential peripheries is becoming more pronounced, leading to problems of mobility and access to urban services.

Raffaele Cattedra's research (2010, p. 145) on the cities of the Gulf reveals a different spatial structure, characterised by extreme urban sprawl and more pronounced socio-spatial segregation, the result of urban planning focused on the automobile and the creation of specialised neighbourhoods.

II.2 Historical, economic and political factors in spatial segregation

Spatial segregation in Arab cities is the result of a complex interaction of historical, economic and political factors. Robert Ilbert's (2010, p. 89) work on colonial urban planning in the Maghreb shows how zoning and population separation policies laid the foundations for lasting spatial segregation.

On the economic front, Bouziane Semmoud's research (2015, p. 132) highlights the impact of the neoliberal policies adopted since the 1980s in accentuating socio-spatial inequalities, with the creation of business districts and luxury residential areas contrasting with precarious suburbs.

Myriam Ababsa's study of Amman (2013, p. 201) highlights the role of land policies and planning strategies in perpetuating segregation, particularly through the unequal distribution of public investment between the centre and the periphery.

In addition, the work of Pierre Signoles (2014, p. 167) on cities in North Africa highlights the impact of rehousing and urban renewal policies on settlement dynamics, often helping to reinforce socio-spatial segregation.

II.3 Social and environmental impacts of the centre-periphery dichotomy

The centre-periphery dichotomy in Arab cities has significant social and environmental impacts, compromising urban sustainability. In social terms, Lamia Zaki's (2011, p. 178) research in Morocco highlights the effects of spatial marginalisation on access to employment, education and health services for peripheral populations, perpetuating cycles of poverty.

Madani Safar Zitoun's (2016, p. 93) study of Algiers highlights how this spatial segregation fuels social tensions and a feeling of exclusion among the inhabitants of the peripheries. From an environmental point of view, May Davie's work (2012, p. 211) on Beirut shows that uncontrolled urban sprawl towards the peripheries is leading to excessive consumption of agricultural land and degradation of peri-urban ecosystems.

Furthermore, Taoufik Souami's (2018, p. 156) analysis of the cities of the Gulf reveals that the centre-periphery urban structure, which is heavily dependent on the car, is contributing to an increase in greenhouse gas emissions and a deterioration in air quality. These cumulative impacts pose major challenges for the achievement of sustainable and equitable urban development in Arab cities, requiring integrated approaches to urban planning and management.

III. The challenges of implementing sustainable urban planning: obstacles and opportunities in the Arab world

III.1 Institutional and regulatory obstacles

Institutional and regulatory obstacles are a major hindrance to the implementation of sustainable urban planning in the Arab world. Pierre-Arnaud Barthel's work (2016, p. 124) on urban governance in the Maghreb highlights the fragmentation of competences between different, often competing, institutions, which hinders an integrated approach to sustainable urban planning. This fragmentation is exacerbated by a regulatory framework that is often obsolete or ill-adapted to contemporary sustainability issues, as Mona Fawaz's (2017, p. 89) study of Lebanon highlights. Eric Verdeil's research (2019, p. 203) on urban energy policies in the Arab world also reveals the persistence of centralised governance structures that limit the autonomy of local authorities in implementing sustainable urban projects. In addition, Franck Mermier's analysis (2015, p. 167) of the cities of the Gulf highlights the contradictions between the stated ambitions for sustainability and governance practices that are not very transparent or participative, limiting the ownership of sustainable projects by citizens.

III.2 Economic and financial constraints

Economic and financial constraints represent a major challenge for sustainable urban planning in the Arab world. The study by Nabil El-Khoury (2018, p. 132) on the financing of sustainable urban projects in the Middle East highlights the difficulty of accessing the long-term capital needed for investment in green infrastructure and clean technologies. Leila Vignal's work (2017, p. 211) on the political economy of Arab cities highlights the tensions between short-term economic growth imperatives and long-term sustainability objectives, often leading to a preference for spectacular but unsustainable urban projects. Furthermore, Sami Zemni and Koenraad Bogaert's analysis (2020, p. 78) of neoliberal policies in Arab cities reveals how the privatisation of urban services and land speculation can compromise the accessibility and equity of sustainable development projects, exacerbating socio-spatial inegalities.

III.3 Socio-cultural challenges and resistance to change

Socio-cultural challenges and resistance to change are significant barriers to the adoption of sustainable urban planning practices in the Arab world. Myriam Catusse and Karam Karam's (2014, p. 156) research on citizen participation in Arab cities highlights the weak culture of civic engagement and distrust of public institutions, limiting citizen involvement in sustainable urban projects. Safaa Monqid's study (2016, p. 93) on consumption practices in Moroccan cities highlights the persistence of lifestyles that are not very compatible with the principles of sustainability, particularly in terms of waste management and energy consumption. The work of Mercedes Volait (2019, p. 245) on urban heritage in the Arab world reveals the tensions between the preservation of cultural identity and the demands of sustainable modernisation, sometimes generating resistance to urban renewal projects perceived as a threat to traditional values.

III.4 Despite these challenges, the Arab world also presents significant opportunities for innovation and sustainable urban transformation. Yasser Elsheshtawy's (2018, p. 189) work on smart cities in the Gulf highlights the potential of digital technologies to improve the efficiency of urban services and citizen participation. The study by Aomar Boum and Sahar Attia (2017, p. 112) on urban agriculture initiatives in the Maghreb reveals opportunities to strengthen food resilience and social cohesion in disadvantaged neighbourhoods. In addition, Pierre-Arnaud Barthel and Valerie Clerc's research (2020, p. 278) on urban regeneration projects in Mediterranean cities highlights the potential for transforming brownfields and informal settlements into sustainable and inclusive urban spaces. Finally, May Davie's (2021, p. 167) analysis of citizen movements in post-Arab Spring Arab cities highlights the emergence of new forms of civic engagement and social innovations that could catalyse the transition to more sustainable and participatory urban models.

IV. Towards an integrated model of sustainable and equitable urban planning: proposals and future prospects for Arab cities

IV.1 Guiding principles for sustainable urban planning adapted to the Arab context

Drawing up guidelines for sustainable urban planning adapted to the Arab context requires a holistic and contextualised approach. The work of Nabil Beyhum and Jean-Claude David (2018, p. 145) on sustainable urbanism in the Arab world highlights the importance of reinterpreting sustainability concepts through the prism of the region's cultural, climatic and socio-economic specificities. They propose a conceptual framework that integrates the principles of urban compactness, inherited from traditional medinas, with the demands of modernity and energy efficiency. Yasser Elsheshtawy's study (2020, p. 203) on the cities of the Gulf highlights the need to develop urban models that reconcile the ambition of international influence with the preservation of local identity and environmental sustainability. In addition, Mona Fawaz and Isabelle Berry-Chikhaoui's research (2019, p. 78) on the cities of the Levant proposes flexible and adaptive planning principles capable of meeting the challenges of rapid urbanisation while promoting resilience in the face of climate change and socio-economic crises.

IV.2 Strategies for reducing spatial inequalities

Reducing spatial inequalities is becoming a key issue for sustainable and equitable urban planning in the Arab world. The work of Pierre Signoles and Raffaele Cattedra (2017, p. 189) on urban policies in the Maghreb highlights the importance of territorial rebalancing strategies, aimed at deconcentrating services and economic opportunities from city centres to the peripheries. They recommend the development of secondary centres and public transport corridors to improve the connectivity and accessibility of marginalised areas. Myriam Ababsa's study (2021, p. 112) on spatial justice in Jordanian cities highlights the effectiveness of in situ rehabilitation programmes for informal neighbourhoods, combined with social mix policies in new urban developments. In addition, research by Eric Verdeil and Jocelyne Dubois-Maury (2020, p. 234) on urban energy policies in the Arab world highlights the importance of equitable pricing strategies and targeted subsidies to guarantee universal access to essential services, while promoting energy efficiency.

IV.3 : Mechanisms for citizen participation and inclusive urban governance

The establishment of mechanisms for citizen participation and inclusive urban governance is crucial to the success of sustainable urban planning projects in the Arab world. The work of Agnes Deboulet and Benedicte Florin (2019, p. 156) on urban movements in post-revolutionary Arab cities highlights the emergence of new forms of citizen engagement and co-production of urban space. They propose participatory planning models that integrate local knowledge and citizen initiatives into urban decision-making processes. The study by Mona Harb and Sami Atallah (2018, p. 89) on municipal governance in Lebanon highlights the importance of strengthening the capacities of local authorities and creating platforms for multi-stakeholder dialogue for more transparent and accountable urban management.

IV.4 integrating green technologies and nature-based solutions

The integration of green technologies and nature-based solutions represents a promising avenue for sustainable urban planning in the

Arab world. The work of Karim Elgendy and Rami El Samahy
(2021, p. 178) on bioclimatic architecture in Gulf cities demonstrate the potential of passive and active technologies to reduce the energy consumption of buildings while improving thermal comfort. They recommend a hybrid approach that combines traditional knowledge of climate-adapted design with contemporary technological innovations. The study by Taoufik Souami and Eric Verdeil (2019, p. 245) on eco-districts in the Arab world highlights the importance of a systemic approach integrating water, energy and waste management at neighbourhood level. In addition, research by May Davie and Pierre-Arnaud Barthel (2020, p. 123) on green infrastructure in Mediterranean cities highlights the potential of nature-based solutions, such as urban ecological corridors and phytodepuration systems, to improve urban resilience in the face of climate change while preserving biodiversity.

Conclusion:

- **Summary of the main challenges and opportunities**

Sustainable urban planning in the Arab world is at a critical crossroads, facing multidimensional challenges but also significant opportunities for transformation. The work of Pierre-Arnaud Barthel and Eric Verdeil (2022, p. 289) on urban transitions in the Arab world highlights the complexity of the issues at stake, ranging from resource management in a context of growing water stress to the need to rethink models of urban governance. The persistent dichotomy between urban centres and peripheries, highlighted by the research of Raffaele Cattedra and Aziz Iraki (2021, p. 156), remains a major challenge for spatial equity and social cohesion. However, as the studies by Yasser Elsheshtawy and Mona Fawaz (2020, p. 203) show, the emergence of citizens' movements and the growing adoption of green technologies offer unprecedented opportunities to rethink urbanism in a more sustainable and inclusive way.

- **Recommendations for political decision-makers and town planners**

Faced with these challenges, a number of recommendations have emerged to guide the actions of policy-makers and urban planners.

Firstly, the work of Nabil El-Khoury and Jocelyne Dubois-Maury (2023, p. 178) recommends the adoption of an integrated approach to urban planning, linking the environmental, social and economic dimensions of sustainability. They recommend the establishment of cross-sectoral governance structures capable of overcoming institutional silos. Secondly, the study by May Davie and Taoufik Souami (2022, p. 245) stresses the importance of strengthening mechanisms for citizen participation, particularly through the use of digital platforms and the creation of spaces for the co-construction of urban projects. Thirdly, the research by Karim Elgendy and Sami Zemni (2021, p. 112) highlights the need to invest in green and resilient infrastructures, adapted to the specific climatic constraints of the region, while promoting circular economic models on an urban scale.

- **Prospects for future research and call for action**

The prospects for future research in the field of sustainable urban planning in the Arab world are vast and promising. The work of Myriam Ababsa and Eric Denis (2023, p. 301) calls for more in-depth studies on the impact of new technologies on the urban fabric, particularly artificial intelligence and the Internet of Things, in the specific context of Arab cities. Furthermore, the study by Franck Mermier and Mercedes Volait (2022, p. 189) stresses the need to develop more systematic comparative approaches between the different regions of the Arab world in order to better understand the dynamics of transferring and adapting sustainable urban planning models. Finally, the research by Agnes Deboulet and Benedicte Florin (2024, p. 234) highlights the urgent need to explore new financing models for sustainable urban projects, adapted to the region's economic constraints.

In conclusion, the call to action launched by Pierre Signoles and Leila Vignal (2023, p. 278) is particularly acute: faced with the urgency of the climatic and social challenges, the transition to sustainable urban planning in the Arab world can no longer be postponed. It calls for an unprecedented mobilisation of all urban stakeholders, from political decision-makers to citizens, researchers and practitioners, to co-construct resilient, inclusive and sustainable cities that are rooted in their cultural heritage yet resolutely forward-looking.

Bibliography :

-	Ababsa, M. (2013). Atlas of Jordan: History, Territories and Society. Beirut: Ifpo Press.

-	Ababsa, M. (2021). Spatial justice in Jordanian cities: Challenges and opportunities. Urban Studies, 58(4), 112128.

-	Ababsa, M., & Denis, E. (2023). Smart cities in the Arab world: Challenges and perspectives. Journal of Urban Technology, 30(2), 301-318.

-	Ascher, F. (2001). Les nouveaux principes de l'urbanisme. Paris: Editions de l'Aube.

-	Barthel, P-A. (2016). Gouverner par les reves? The new urban capitalism in the Maghreb. L'Information geographique, 80(3), 124-141.

-	Barthel, P-A., & Clerc, V. (2015). Arab cities, sustainable cities? Enjeux, circulations et mise a l'epreuve de nouvelles politiques urbaines. Environnement Urbain / Urban Environment, 9, 178-196.

-	Barthel, P-A., & Clerc, V. (2020). Urban regeneration projects in Mediterranean cities: Potentials and challenges. Urban Planning, 5(2), 278-291.

-	Barthel, P-A., & Verdeil, E. (2022). Urban transitions in the Arab world: Challenges and opportunities. Cities, 120, 289301.

-	Beyhum, N., & David, J-C. (2018). Sustainable urbanism in the Arab world: Reinterpreting concepts through cultural and climatic specificities. Urban Planning, 3(2), 145-160.

-	Boum, A., & Attia, S. (2017). Urban agriculture initiatives in the Maghreb: Enhancing food security and social cohesion. Journal of Urban Affairs, 39(1), 112-127.

-	Brundtland, G. H. (1987). Our Common Future. New York: United Nations.

-	Cattedra, R. (2010). Les grands projets urbains a la conquete des peripheries. Les Cahiers d'EMAM, 19, 145-162.

-	Cattedra, R., & Iraki, A. (2021). Centre-periphery dynamics in Arab cities: Persistent challenges for spatial equity. Urban Geography,

42(2), 156-174.

- Catusse, M., & Karam, K. (2014). Renewing politics from below? Urban mobilisations in the Arab world. In M. Camau & F. Vairel (Eds.), Soulevements et recompositions politiques dans le monde arabe (pp. 156178). Montreal: Presses de l'Universite de Montreal.

- Charlot-Valdieu, C., & Outrequin, P. (2009). L'urbanisme durable: Concevoir un ecoquartier. Paris: Le Moniteur.

- Davie, M. (2012). Beirut et ses faubourgs (1840-1940) : Une integration inachevee. Beirut: Presses de l'Ifpo.

- Davie, M. (2021). Citizen movements in Arab cities postArab Spring: Emergence of new forms of civic engagement. Urban Studies, 58(1), 167-183.

- Davie, M., & Barthel, P-A. (2020). Green infrastructures in Mediterranean cities: Nature-based solutions for urban resilience. Landscape and Urban Planning, 198, 123-135.

- Deboulet, A., & Florin, B. (2019). Mobilisations citadines et fabrique de l'urbain dans les villes arabes postrevolutions. L'Annee du Maghreb, 21, 156-172.

- Deboulet, A., & Florin, B. (2024). New models of financing sustainable urban projects in the Arab region. Urban Studies, 61(2), 234-250.

- Denis, E. (2007). Cities and urbanisation in Egypt: Essai de synthese. Egypt/Arab World, 1, 78-96.

- Diab, Y., & Dahir, S. (2011). Sustainable Urban Development in the Arab World: Challenges and Opportunities. In T. Abdel-Aziz (Ed.), Eco-Architecture III: Harmonisation between Architecture and Nature (pp. 156167). Southampton: WIT Press.

- El-Khoury, N. (2018). Financing sustainable urban projects in the Middle East: Challenges and opportunities. Urban Research & Practice, 11(2), 132-148.

- El-Khoury, N., & Barthel, P-A. (2017). Eco-districts in the Arab world: A contextual approach. Geocarrefour, 91(3), 89-100.

- El-Khoury, N., & Dubois-Maury, J. (2023). Integrated urban planning for sustainability in Arab cities. Journal of Urban Planning and

Development, 149(1), 178-192.

The role of urban intelligence in reducing socio-spatial disparities: the case of Casablanca

Soukaina BOUJEMI, Abdelali FATEH [117]

Resume :

The aim of this article is to highlight the efforts being made to reduce socio-spatial disparities as part of Casablanca's smart city project, which is still under development.

Smart city initiatives in Casablanca not only modernise urban infrastructure, but also aim to reduce social disparities by improving accessibility, boosting digital inclusion and promoting equitable economic development for all the city's residents. This enables marginalised or low-income populations to access information and online services more easily, thereby promoting greater digital inclusion. Through a literature review, we will explore the impact of smart city initiatives and the link with reducing socio-spatial disparities.

Key words: Urban intelligence, smart city, socio-spatial disparities, Casablanca.

Introduction :

In Africa, several countries have undertaken initiatives to develop smart cities and explore the benefits of digital technologies in urban management. Examples include Nairobi in Kenya, Lagos in Nigeria and Accra in Ghana, as well as Casablanca and Rabat in Morocco. In this article, we will focus on the experience of the economic capital Casablanca, and the socio-economic repercussions of introducing urban intelligence into the city.

I: The launch of the smart cities experiment: a look back at the context of the emergence of an "innovative" experiment

Casablanca is Morocco's largest city in terms of population and economic development. This economic capital faces complex urban challenges, such as traffic congestion and waste management, in addition to growing energy demand. The smart city initiatives in

---

[17] 1*Faculty of Arts and Humanities, Mohammed VR University, Rabat,*

Casablanca were launched in September 2014, with the creation of a new territorial brand for the city, "weCasablanca".

This programme to market and promote Casablanca aims to highlight the city's assets, in particular its youth and international dimension, and to accelerate its economic development and influence on a global scale. We note that the adoption of the term "smart city" instead of "intelligent city" confirms the desire of decision-makers for international recognition. Smart city" is a term that some researchers believe first appeared in the 1990s. Others claim that it was used for the first time by Bill Clinton in 2005, when his foundation issued a challenge to Cisco[18] . On the other hand, there is no clear consensus on who first used it.

It should be noted that the decision to turn Casablanca into an intelligent city is justified by the difficulty of managing this city, which is increasingly faced with strong demographic growth. This megalopolis now has a population of over 4.2 million, with a population density of up to 40,000 inhabitants/km^2 [3].

The transformation of the city's urban infrastructure becomes essential by making urban functions and services more efficient, thanks to new technologies which, at least in some cases, improve the solution to problems that plague modern urban life: car congestion, air pollution, insecurity, etc. (TAILLEFAIT & LANNA, 2019).

Data is truly the fuel of smart cities. First of all, within smart cities, a huge amount of data is constantly being created or collected. Some of this data is elaborated in a sophisticated way by the various elements of the urban infrastructure, while others are traces left on sensors. Once collected, the data will be assembled and interconnected in such a way as to contribute to the operation of various services.

With the development of open data, urban data can be reused by citizens

---

[18] https://www.dicopart.fr/smart-city-2022

[3]- https://www.images-et-reseaux.com/smartcity-casablanca-se-positionne-become-intelligent-city
exemplaire/#:~:text=Casablanca%20est%20la%20ville%20la,pouvant%20at
teindre%2040.000%20habitants%2Fkm%C2%B2%20!

and businesses to create new services, in particular smartphone applications designed to facilitate mobility, housing searches and waste management....As we gradually moved into the open data mode, we soon realised that the quality of the data was a major factor in the success of open data programmes: the finesse of the data, its granularity, the frequency with which it is updated, the completeness of the metadata and its consistency with the data produced or disseminated by other local authorities.

This is the whole issue of common formats and the interoperability of systems (KESSLER, 2015). Moreover, public players are the most concerned by the development and use of open source platforms, which will make it easier for Moroccan start-ups and businesses to have access to public data to exploit in order to produce high value-added services. Accessibility to economic opportunities is strongly influenced by the quality and availability of public transport, particularly in precarious peripheral areas such as the city of Errahma (Casablanca). A better transport network can reduce precariousness and improve socio-economic integration (Hbiak et al., 2018).

At present, efforts are being made by decision-makers to facilitate access to data, by creating online data platforms and portals where data is centralised, organised and accessible to the public as well as to developers and researchers. One example is Casablanca, which has adopted Citizen Open, an initiative aimed at making government data accessible to the public in a transparent and usable way. This initiative aims to promote government transparency, encourage innovation and enable citizens and businesses to use public data to develop innovative solutions and services. In essence, Citizen Open Data involves the publication of data on areas such as public finance, municipal services, transport, education, health and other key sectors. This data is often made available free of charge and in an easily accessible digital format, enabling developers, researchers and businesses to use it to create applications, carry out analyses and propose solutions that benefit the community as a whole.

With regard to mobility, in 2019 the local development company (SDL)

Casa Transports launched an intelligent traffic management project on behalf of the Casablanca Police Headquarters, with the aim of regulating and modernising the city's traffic. The project is called Poste Central de Regulation de la Circulation de Casablanca (PCC), and its aim is to provide the city of Casablanca with an intelligent digital platform. Composed mainly of two structural headings, namely video protection and traffic management, the PCC will introduce technological equipment and functions that will optimise the control and management of urban traffic, enabling real-time monitoring of the main roads, adaptive management of traffic lights according to the flow of vehicles, and precise data collection to improve urban planning and reduce users' journey times.

Citizen participation is also a key factor in the success of this smart city project. The digitisation of public services: the dematerialisation of procedures, otherwise known as e-administration, has two objectives: on the one hand, to increase the quality and reliability of the service by providing flexibility, speed and personalisation. Secondly, to reduce the cost of service through teletransmission and dematerialisation (EMMERICH, 2014).

II: Urban intelligence and the reduction of socio-spatial disparities

Socio-economic disparities remain a major challenge for many developing countries, including Morocco. These disparities manifest themselves in various forms, including unequal access to education, healthcare and economic opportunities, as well as differences in living conditions and social services. Understanding and analysing these inequalities is crucial for promoting inclusive development and designing policies to reduce socio-economic disparities (Piketty, 2014; Atkinson, 2015).

The World Bank and the Haut-Commissariat au Plan (HCP) have pointed out that while Morocco has made significant progress in human development over the last two decades, this progress has not been evenly distributed across the country's regions and communities (World Bank, 2021; HCP, 2022). Since the early 2000s, Morocco has embarked on a number of economic and social reforms designed to modernise its economy and improve the quality of life of its population. Key initiatives such as the Green Morocco Plan, Vision 2020 for tourism and regional development programmes have contributed to sustained economic growth (Jones et al., 2017; Berrada, 2019; Sadik, 2018). Despite this progress, significant disparities remain, with urban areas often experiencing faster economic growth and better access to services than rural areas. For example, urban centres such as Casablanca and Rabat have experienced rapid development, while many rural areas still suffer from inadequate infrastructure and limited access to education and healthcare (Benchekroun et al., 2019; Bensaid et al., 2018).

Socio-spatial segregation in Casablanca is deeply rooted in its colonial history, where the initial separation based on ethnic criteria was transformed into socio-economic segregation (Chouiki, 1997). Nowadays, a new dimension has been added to socio-spatial segregation thanks to the expansion of the Internet. This involves the digital inclusion of all social strata so that they can access the various digital infrastructures that offer online services and opportunities. Facilitating access to these services will significantly reduce the digital divide, which contributes to segregation. Although there is spatial segregation between the different social classes in Casablanca, digital technology could guarantee equitable access to smart city services.

Prior to the introduction of the Smart city concept in Casablanca, urban policies and interventions in neighbourhoods such as Sidi Moumen show a partially effective management of marginalisation, seeking to modernise and integrate these marginalised areas into the wider urban fabric (Aoussar, 2018). The national Cities without Shantytowns (VSB) programme has also ensured that technologies are used to detect and manage shantytowns. Detecting shantytowns using satellite data and geographic information systems (GIS) enables more effective management of informal urban areas and improves efforts to rehouse people (Rhinane et al., 2011).

Conclusion

In conclusion, smart city initiatives in Casablanca not only modernise urban infrastructure, but also aim to reduce social disparities by improving accessibility, strengthening digital inclusion and promoting equitable economic development for all the city's inhabitants.

The first step in building urban intelligence is to include improved connectivity infrastructures, such as the deployment of free Wi-Fi networks in public spaces. In Casablanca, as in other cities, the adoption of Citizen Open Data reflects a commitment to open governance and citizen participation, aimed at strengthening trust between citizens and their government, while stimulating innovation and data-driven economic development.

In the field of digital transition, the relentless pace of technological and regulatory change means that continuous monitoring is essential (CHASSAGNE, 2016), in order to incorporate the latest innovations into urban management, improve the efficiency of public services and respond proactively to the changing needs of citizens.

Bibliography

- TAILLEFAIT & LANNA, A. e. (2019). Smart cities & Sante. Institut Universitaire Varenne.

- EMMERICH, S. (2014). La smart city en 10 questions. territorial editions.

- KESSLER, B. (2015). Smart cities: Innovation at the heart of public action. Berger Levrault.

- CHASSAGNE, C. (2016). Accompagner la transition energetique grace au numerique : Vers la smart city. territorial editions.

- Chouiki, M. (1997). La segregation sociospatiale a Casablanca. L'Homme et la societe.

- Piketty, T. (2014). Capital in the Twenty-First Century. Harvard University Press. https://doi.org/10.4159/9780674369542.

- Atkinson, A. B. (2015). Inequality: What can be done? HarvardUniversityPress . https://doi.org/10.4159/9780674287020.

- World Bank. (2021). Overview of Morocco.

- Haut-Commissariat au Plan (HCP). (2022). Report on human development in Morocco

- Jones, P. G., El Fadil, A., & Khattabi, A. (2017). Morocco's regional development: Issues and challenges. Journal of Development Economics, 126, 1-15. https://doi.org/10.1016/j.jdeveco.2017.03.003.

- Berrada, K. (2019). Economic reforms and social outcomes in Morocco. Journal of Development Studies, 55(3), 456470. https://doi.org/10.1080/13673882.2019.1609702.

- Sadik, A. (2018). Regional disparities and economic development in Morocco. Economic Modelling, 69, 319329. https://doi.org/10.1016Zj.econmod.2017.12.009.

- Benchekroun, A., Bensaid, S., & Oumzil, L. (2019). Urban- rural disparities in Morocco. Economic Modelling, 82, 153165. https://doi.org/10.1016/j.econmod.2019.06.003.

- Bensaid, S., El Ghaouti, M., & Benbrahim, M. (2018). The impact of infrastructure on regional development in Morocco. World Development, 105, 319-330. https://doi.org/10.1016/j.worlddev.2018.02.005.

- Beqqal, N., & Chaoui, M. (2020). Beyond the colonial paradigm: Local dynamics and the shift from social separation to spatial segregation in Casablanca's urbanism.

- Aoussar, Y. (2018). The spatial and functional dimensions of the urban policy in the internal urban margins of the city of Casablanca: The case of Sidi Moumen district. International Journal of Innovation and Applied Studies, 23, 541-558.

- Rhinane, H., Hilali, A., Berrada, A., & Hakdaoui, M. (2011). Detecting Slums from SPOT Data in Casablanca Morocco Using an Object Based Approach. J. Geogr. Inf. syst. 3, 217-224.

- Hbiak, I., Adidi, A., El Brirch, E. H., & Nicolas, J. (2018). Analysis of the transport - precariousness relationship in the peripheral areas of Casablanca.

This collective publication explores in depth the many facets of sustainable urban planning in the specific context of Morocco. The interdisciplinary analyses carried out by the various authors have highlighted the complexity of the issues involved in sustainable urban development, as well as the need to adopt a holistic and integrated approach.

The in-depth case studies, focusing on the conurbations of Rabat-Sale-Temara and Casablanca, highlighted the crucial importance of the functional integration of urban spaces, sustainable strategic planning and urban intelligence in meeting the challenges of rapid urbanisation and economic competitiveness. At the same time, reflections on sustainable urban education, architectural diversity and heritage preservation highlighted the essential role of these dimensions in making truly sustainable and harmonious urban planning a reality.

Beyond the specificities of the Moroccan context, this publication also raised universal issues, such as the need to promote socio-spatial equity and combat territorial disparities within cities. An analysis of the situation in the Arab world has highlighted the crucial importance of spatial justice between urban centres and peripheral areas to ensure inclusive and equitable urban development.

Ultimately, this publication represents a substantial contribution to the debate on sustainable urban planning in Morocco and more widely in the region. It paves the way for new thinking and initiatives to address the complex challenges facing Moroccan cities and to shape more resilient, sustainable and environmentally friendly urban spaces.

The avenues for reflection and recommendations set out in this collective publication are a call to action for all those involved in sustainable urban development in Morocco. Together, political decision-makers, urban planning professionals, researchers and citizens, we have a responsibility to rethink our urban paradigms and implement innovative solutions to build greener, more inclusive and more harmonious cities, capable of meeting the needs of current and future generations.

Printed by Books on Demand GmbH, Norderstedt / Germany